Spaceships

A Reference Guide to International Reusable
Launch Vehicle Concepts from 1944 to Present

Robert A. Goehlich

An Apogee Books Publication

Published by Apogee Books, Box 62034, Burlington,
Ontario, Canada, L7R 4K2, http://www.apogeebooks.com
Tel: 905 637 5737

Printed and bound in Canada
Spaceships by Robert Goehlich
ISBN 978-1-894959-50-6
©2005 Apogee Books

Abstract

Space transportation is one of the most essential elements for enabling activities in space. For current rockets, reliability is too low and launch cost is too high when compared to aircraft operations. Reusable Launch Vehicles could solve these deficiencies and are being investigated by many companies. This book contains a databank of 300 worldwide suborbital and orbital Reusable Launch Vehicle concepts. It covers ideas from the first concepts, such as Silver Bird, proposed by Eugen Saenger in 1944, to present ones such as SpaceShipOne, proposed by Burt Rutan in 2003, as well as all X Prize candidates. For reader friendly use, all information is prepared in the same data style, which makes this book a unique reference for rocket scientists as well as everybody interested in and fascinated by rockets. An introduction to space transportation systems, a study on the motivation for developing Reusable Launch Vehicles and a discussion about the benefit of an international Reusable Launch Vehicle program complete this book.

KEYWORDS: Orbital Flight, Reusable Launch Vehicle, Space Transportation System, Space Tourism, Suborbital Flight

STATISTICS: 132 pages, 264 figures, 5 tables, 22 references

DATA COLLECTION: Robert A. Goehlich, JAXA / Keio University, Japan
                       Masayoshi Ogawa, Fuji Heavy Industries, Japan
                       Yasufumi Ogasawara, Canon Inc., Japan

CONTACT: robert@goehlich.com; www.goehlich.com

Acknowledgements

The realization of the present book would not have been possible without the contribution of many other persons. They all spared some of their time making themselves available for personal communications or provided documents that would otherwise not have been available. I would like to take this opportunity to express my appreciation of this support to:

A.C. Charania, SpaceWorks Engineering Inc.

Dominique Demougin, Business and Economics Faculty, Humboldt University at Berlin

Rob Godwin, CG Publishing Inc.

Lothar and Rosemarie Goehlich, Institute for High Voltage Engineering, Technical University Berlin

Heinz-Hermann Koelle, Institute of Aero- and Astronautics, Technical University Berlin

Naoko Ogawa, International Passenger Revenue Accounting Center, Japan Airlines

Yoshiaki Ohkami, Tsukuba Space Center, JAXA

Hirokazu Suzuki, Future Space Transportation Research Center, JAXA

Kazuo Yoshida, Faculty of Science &Technology, Keio University

Rodney VanMeter, Faculty of Science &Technology, Keio University

The RLV concept study project and the publication of the results in the form of this book was funded by the Japan Society for the Promotion of Science (JSPS) and the Alexander von Humboldt Foundation (AvH Foundation), which is gratefully acknowledged by the author.

Preface

The aim of this book is to provide the general public as well as aerospace engineers with detailed information about the outstanding variety of smart rocket concepts. For reader friendly use, all information is prepared in the same data style. The content of this book consists of the following two parts: Part 1 provides an introduction of the general characteristics of Space Transportation Systems and Reusable Launch Vehicles. Part 2 is the backbone of this book, consisting of a databank of more than 300 worldwide Reusable Launch Vehicle concepts.

Space transportation is one of the most essential elements for enabling activities in space. For current rockets, reliability is too low and launch cost is too high when compared to aircraft operations. Reusable Launch Vehicles (RLVs) could solve these deficiencies and are being investigated by many companies. RLVs are designed for quick-turnaround operations that will allow for a higher volume and launch rate, resulting in economies of scale. Assets of RLVs are low operating costs for high launch rates, high reliability and satisfactory ecological compatibility. Known disadvantages of RLVs are high development costs and high operating costs for low launch rates similar to Space Shuttle system operations. RLV concepts proposed for development present a variety of launch, landing and propulsion concepts. Several vehicles employ a spaceplane design that might take off and land horizontally like an airplane, possibly incorporating air-breathing engines for atmospheric flight. These designs generally use upper stages to carry payloads to orbit, while the spaceplane remains on a suborbital trajectory. Many of these vehicle concepts are conceived with the expectation that there will be significant demand for launches of communication satellites, some hope to serve other new markets such as space station resupply and flights for space tourists.

As you can figure out from this book, there are many similarities of rocket designs between countries and also repeated over the time. I hope by structuring RLV concepts as it is done in this book, it may help you to push your RLV research a little bit further.

If your RLV project is not listed or incompletely listed, please complete the blank form from www.goehlich.com at publications section and send it to robert@goehlich.com. It will be added for the next edition of "Spaceships".

Yokohama, September 2005   Robert A. Goehlich

Table of Contents

List of Figures

List of Tables

List of Abbreviations

-e	[-]	expendable
-r	[-]	reusable
AAS	[-]	Ascent / Accelerate Support
AE	[-]	Air-breathing Engine
ARG	[-]	Argentina
BLN	[-]	Balloon
CHN	[-]	China
CNES	[-]	Centre National d'Etudes Spatiales
deg	[-]	degree
ESA	[-]	European Space Agency
FBB	[-]	Fly-back Booster
FRA	[-]	France
GBR	[-]	United Kingdom
GEO	[-]	Geostationary Orbit
GER	[-]	Germany
HTHL	[-]	Horizontal Take-off / Horizontal Landing
HTVL	[-]	Horizontal Take-off / Vertical Landing
IND	[-]	India
ISR	[-]	Israel
ISS	[-]	International Space Station
JAXA	[-]	Japan Aerospace Exploration Agency
JPN	[-]	Japan
km	[-]	kilometer
LEO	[-]	Low Earth Orbit
Mg	[-]	Mega grams
NASA	[-]	National Aeronautics and Space Administration
ORB	[-]	Orbiter
PLN	[-]	Aero Plane
RBCC	[-]	Rocket Based Combined Cycle
RE	[-]	Rocket Engine
RKT	[-]	Rocket Stage
RLV	[-]	Reusable Launch Vehicle
ROM	[-]	Romania
RUS	[-]	Russia
s	[-]	second
SRB	[-]	Solid Rocket Booster
SSTO	[-]	Single Stage to Orbit
TSTO	[-]	Two Stage to Orbit
VTHL	[-]	Vertical Take-off / Horizontal Landing
VTVL	[-]	Vertical Take-off / Vertical Landing

Definitions

This section provides a description of all items used in the spaceship's data sheets as shown in Figure 1. In general, the data sheets are alphabetically sorted by country and within each country they are alphabetically sorted by vehicle names.

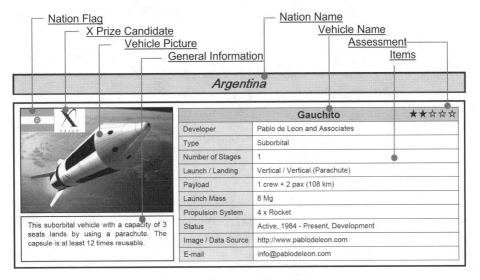

Figure 1: Example of Data Style

Vehicle Name   "Vehicle Name" is basically the name of the orbiter. Since RLV systems often consist of some parts such as booster and orbiter, "Vehicle Name" can indicate the name of the total system of the RLV concept. For example, Space Shuttle is the name of the system that includes the orbiter, solid rocket boosters and external tank. In this book, when there is a proper name for each stage vehicle, they are shown in round brackets, for example "SpaceShipOne (Booster: White Knight, Orbiter: SpaceShipOne)". In this case SpaceShipOne is the orbiter and White Knight is used on the way to space to carry the orbiter. In the datasheet, "Booster" is used as the name for the lower stage, and its function is to accelerate the velocity or increase the altitude of the vehicle. "Orbiter" is the part that reaches the target orbit.

Assessment   Judgment is based on the estimated probability for technical, economic and political feasibility. The judgment is based on data available to the RLV concept project team at Keio University. The range of scale is from 0 to 5 stars. 0 stars means low probability for feasibility, 3 stars means neutral probability for feasibility, while 5 stars means high probability for feasibility.

Developer   Developer can be a company, a university or an individual.

Type   "Type" shows the objective orbit of proposed concept. Basically, objective orbits are defined as "Suborbital" or "Orbital". A suborbital vehicle is expected to reach about 100 km altitude. When the objective orbit is called Orbital, it has an altitude like Low Earth Orbit (LEO) or Geostationary Orbit (GEO). LEO is an abbreviation for Low Earth Orbit. This term is typically used to describe the orbital altitude range (300 km to 2000 km above the surface of the Earth) of certain communications satellites (ANSI, 2000a). GEO is a circular orbit in the equatorial plane, any point on which revolves about the Earth in the same direction and with the same period as the Earth's rotation. Its altitude is approximately 35 786 km above mean sea level (ANSI, 2000b). For example the orbit of the International Space Station ISS is a circular one whose altitude is from 330 km to 480 km, so it is a LEO.

Number of Stages   "Number of Stages" is number of stages comprising the proposed system. The technique of staging involves dropping off fuel tanks and structures, including engines, when the fuel is burned off (Olivier, 1994). In this book, one stage is defined as the vehicle part that can increase altitude or velocity by its force. Sometimes an orbiter does not satisfy the definition, but an orbiter is always counted as one stage. For example, the Space Shuttle's number of stages is 2. It consists of the orbiter, solid rocket boosters and external tank. The external tank has no force to increase altitude or velocity, while the orbiter and solid rocket boosters fulfill the definition for a stage. In addition to stage number, this book shows details of each stage in brackets. In brackets, there are 6 kinds of stages that are "RKT", "FBB", "PLN", "AAS", "BLN" and "ORB":

- RKT means RocKeT Stage, which is a stage for acceleration or ascent without aerodynamic force but rocket engine force. RKT includes a vertical ascent rocket stage, which is either expendable or reusable. The Space Shuttle's SRB is classified as RKT.

- FBB means Fly-Back Booster. During ascent, a fly-back booster's function is same as RKT. The difference between FBB and RKT is that FBB has its own wing and it can fly back to the launch site by using aerodynamic forces.
- PLN means aero PLaNe. An aero plane stage is a stage for acceleration or ascent with aerodynamic force like an aircraft.
- AAS is an abbreviation for Ascent / Accelerate Support. ASS is used when the stage cannot be classified as RKT, FBB or PLN.
- BLN means BaLlooN that lift up the vehicle by its lift force.
- ORB means ORBiter.

Each abbreviation may have a suffix "-r" or "-e", if the information is available. "-r" means reusable and "-e" means expendable. For example "RKT-r: 1, ORB-r: 1" means: "This concept has 1 reusable rocket stage and 1 reusable orbiter stage".

Launch / Landing   "Launch / Landing" shows how the proposed system takes off or lands. Generally, Space Transportation Systems have two styles of launching and landing. They are "Horizontal" and "Vertical". At present, most transportation systems use the vertical launch style. For example, the Space Shuttle is launched vertically and its landing style is horizontal like an aircraft.

Payload   There are two kinds of payloads, which are persons and cargo. "Payload" is the goods carried by a vehicle. In this book, persons are classified into passengers and crew, while cargo is expressed by its mass. As performance of the transportation system is much influenced by the latitude of launch site, the author added the latitude of launch site and objective altitude in round brackets if this information is available. For example "Payload" of Space Shuttle is shown as "7 crew + 25,0 Mg (28 deg / 204 km)". The mission of it is not commercial so there is no passenger.

Launch Mass   "Launch Mass" is the mass of a RLV launched from the ground. "Launch Mass" can also be called take-off weight/mass, total weight/mass or launch weight. Generally, "Launch Mass" consists of structural mass, propulsion system mass, equipment mass, payload mass and propellant mass (Harloff, 1988). The structural mass includes vehicle's body and tanks, wings, thermal protection systems and gear. The equipment mass consists of hydraulic systems, avionics and electronic systems. For example, Space Shuttle's launch mass is about 2035 Mg. It consists of 2 solid rocket boosters (1180 Mg), 1 external tank (751 Mg) and 1 orbiter (104 Mg), which contains the payload.

Propulsion System In this book, "Propulsion System" indicates an engine of the vehicle that generates thrust. In a general definition, the propulsion system is not only an engine but also the structures holding the engine, pipes supplying propellant to the engine and so on. There are several kinds of propulsion systems, for example ion, nuclear and chemical powered engines. Today, mainly chemical engines are used. They are categorized into two types, air-breathing engines and rocket engines. The definition of air-breathing engine includes also the category of jet engines used for aircraft. The most important difference between them is that an air-breathing engine uses oxygen from the atmosphere, while a rocket engine carries oxygen in the vehicle's tank. For example, the Space Shuttle uses only rocket engines.

Status "Status" indicates the phase of development. The first clause is "Active" or "Inactive". Active means that the vehicle concept is now under study, in development or in operation. Inactive means that the project of the proposed concept has been finished or development has been stopped. The second clause is about the period of the project. The period description is for example indicated as "1990 - 1998", meaning that the project was started in 1990 and was stopped in 1998. The third clause provides some detail. "Concept" is the phase of concept study. In this phase developers consider the configuration or calculate the feasibility. "Development" means that they build some experimental model or sub systems like engines, thermal protection systems, etc. "Realized" indicates that the concept has been realized. For instance, the Space Shuttle is "Realized" and "Active".

Image / Data Source This homepage address gives the reference from which the images and data have been taken. High quality images and detailed vehicle specifications can be found here.

E-mail To contact the developer directly, please use this e-mail.

Part 1

About Spaceships

1 PART I: About Spaceships

After a brief introduction to space transportation systems, this chapter summarizes the justification to investigate the potential of Reusable Launch Vehicles (RLVs) based on the current and foreseeable state-of-the-art. Aerospace experts may like to skip this very basic information about spaceships. The chapter is concluded with a discussion about the advantages and disadvantages for an international RLV program.

1.1 Space Transportation Systems

In this section a general overview of Space Transportation Systems is given. Space Transportation Systems can be Reusable or Expendable Launch Vehicles. The term Space Transportation Systems refers not only to the vehicle itself but also includes normally the infrastructure such as launch, maintenance and propellant facilities.

1.1.1 Use

Many people think that the most difficult problem to going into space is reaching a high altitude. But to tell the truth, an altitude from 100 km to 400 km is not too hard to reach. If you do not need to stay in orbit, you can arrive there with a speed of about Mach 6 in the vertical direction, which is technically relatively easy to achieve, compared to orbital speed. The most important use of Space Transportation Systems is obtaining the velocity to go around the Earth like a satellite. This is about 8 km/s or Mach 26 as shown in Figure 2.

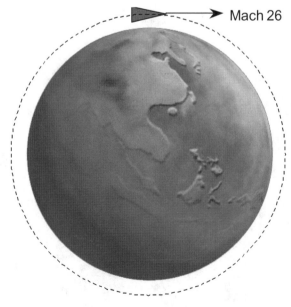

Figure 2: Orbital Flight Velocity

Except for Space Transportation Systems, there are no vehicles that can travel with such very high velocity. For example, the Concorde only reached around Mach 2. Figure 3 illustrates a speed comparison of some vehicles. Gravity, aerodynamic drag forces, etc. decelerate the Space Transportation System. This is called "velocity loss", which is generally about 2 km/s. To compensate this velocity loss the Space Transportation System needs energy which is equivalent to about 10 km/s or Mach 30.

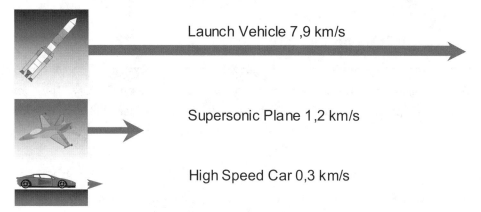

Figure 3: Velocity Comparison (Ogasawara, 2004)

1.1.2 Performance

The choice of the Space Transportation System type is mainly decided by the weight of its payload and the objective orbit condition. The heavier the payload is and the higher the orbit in which that payload will be placed is, the more performance is needed. Some operational vehicles are shown in Figure 4.

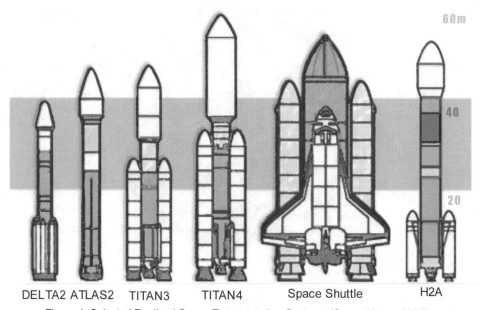

Figure 4: Selected Realized Space Transportation Systems (Space Notes, 2005)

To design a highly functional Space Transportation System, that is, a rocket that works efficiently, the Space Transportation System's body should be as light as possible and its propulsion - the force that propels the rocket - as great as possible (Space Notes, 2005). In addition, the launch site latitude has much influence on the performance of the Space Transportation System. At lower latitudes the performance is higher for the same rocket type due to the Earth rotation. Table 1 shows a selection of realized Space Transportation System performances.

Table 1: Selection of Realized Space Transportation System Performances

Launch Vehicle	Total Mass	Latitude	Payload LEO	Payload GEO
Ariane 5 (ESA)	725 Mg	7,0 deg	22,5 Mg	6,8 Mg
CZ 3A (China)	240 Mg	28,5 deg	7,2 Mg	2,6 Mg
H-IIA (Japan)	285 Mg	30,0 deg	10,0 Mg	4,1 Mg
Soyuz (Russia)	310 Mg	51,8 deg	7,8 Mg	n.a.
Space Shuttle (USA)	2030 Mg	28,5 deg	24,4 Mg	n.a.

1.1.3 Requirements

When the Space Transportation System is traveling to orbit, it uses propellant that con-sists of fuel and oxidizer. They may be stored as a solid, liquid or cryogenic liquid; for example, hydrogen (fuel) is stored in a cryogenic liquid propellant tank. There is a big difference between Space Transportation Systems and ordinary vehicles: the need for an "oxidizer". Cars, motorcycles or airplanes do not need oxidizer tanks, since there is enough oxygen in the air surrounding them.

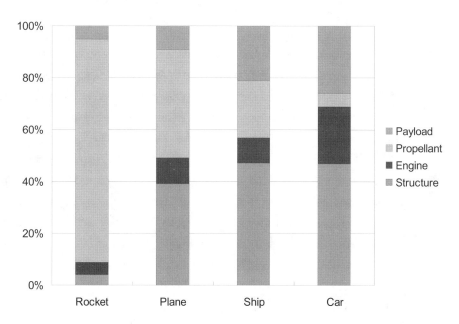

Figure 5: Comparison of Mass-ratios of Selected Vehicles

However, a Space Transportation System has to carry oxidizer in its own body, because there is no air in space. The total mass of a Space Transportation System can be divided into about 80 - 90 % propellant and about 10 - 20 % structures, equipment and payloads (HyperDic, 2005). Figure 5 shows general mass ratios of selected vehicles.

Despite the lightweight bodies, the Space Transportation System must endure the aerodynamic pressure (in the atmosphere), the vibration, the acceleration and the radiation from the sun as shown in Figure 6. These topics have to be considered carefully when the Space Transportation System is designed. The important challenge is how to ensure the safety of payload and vehicle. (Ogasawara, 2004)

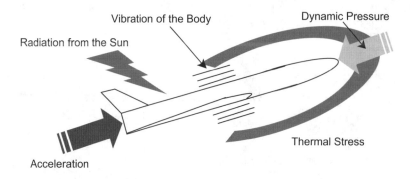

Figure 6: Constraints of the Space Transportation System (Ogawa, 2003)

1.1.4 Materials

To achieve a high performance Space Transportation System, one way is by reducing the mass of the vehicle. Many kinds of materials are used to make the vehicle lightweight. In the case of the H-IIA rocket, aluminum alloys are used for the fairing as well as for propellant tanks, while carbon compound materials are used for the interstages of the rocket. Depending on the item, carbon fiber reinforced plastics, titanium alloys, graphite or Kevlar are also used. For engines, precision processing is required, and aluminum alloys are usually used to cope with the high temperatures (about 3000 degrees Celsius), extremely low temperatures (liquid hydrogen has a temperature of minus 250 degrees Celsius) and the harsh vibrations. Important factors that influence which materials are used include not only the characteristics of individual materials but also the balance of development and production funds. (Space Notes, 2005)

1.1.5 Concept Choice

When you buy a car you can choose some of its characteristics. They are "body color" (black, red, blue), "chassis" (2 wheel drive or 4 wheel drive), "type" (small car, sedan, sports car), etc. Similar to a car, there exist also some main elementary characteristics of Space Transportation Systems. They are "stage type", "propulsion type" and "launch / landing type" to name a few as shown in Table 2.

Table 2: Morphological Box of selected typical Rocket Characteristics

Design Features	Choice of Characteristics			
Stage Type	Single Stage to Orbit (SSTO)		Two Stage to Orbit (TSTO)	
Propulsion Type	Rocket Engine (RE)	Rocket Based Combined Cycle (RBCC)	Air-breathing Engine with Rocket Engine (AE + RE)	
Launch / Landing Type	Horizontal Take-off / Horizontal Landing (HTHL)	Horizontal Take-off / Vertical Landing (HTVL)	Vertical Take-off / Horizontal Landing (VTHL)	Vertical Take-off / Vertical Landing (VTVL)
...	...	...	...	...

A box listing typical alternative characteristics available for each design feature is also called a morphological box (Zwicky, 1966). This box can be used for deriving system-atically promising vehicle concepts. There are many combinations possible that lead to vehicle concepts of different quality concerning technical, economic and political feasi-bility.

1.2 Reusable Launch Vehicle

After describing Space Transportation Systems in general in the previous section, this section investigates in Reusable Launch Vehicles, which represent one important group of Space Transportation Systems.

1.2.1 Differences between Reusable and Expendable Launch Vehicles

Reusable Launch Systems are almost the same as Expendable Launch Systems, but parts of the system must return to Earth in good enough condition to be reused. Ex-pendable Launch Systems basically consist of structures, propulsion system, propellant tanks, avionics and electrical system. The avionics and electrical system is needed to control all subsystems of the Space Transportation System and maneuver the vehicle. In addition to them, RLVs need a Thermal Protection System to resist high temperature friction of air during reentry. If RLVs employ horizontal landing style, they must have wings, vertical / horizontal fins and landing gears. Vertical landing RLVs must have parachute or booster rockets to decelerate.

1.2.2 Reusability

A Reusable Launch Vehicle may be either fully or partly reusable. Except for the pro-pellant tank that is called the External Tank, the Space Shuttle system consists of a reusable orbiter and reusable Solid Rocket Boosters (SRBs). After the flight, the orbiter is serviced to prepare it for the next flight. The two SRB are separated shortly after launch, then float to the sea by parachute. They are retrieved, serviced and used again (Space Notes, 2005), while the External Tank is discarded each flight. Thus, the Space Shuttle system can be called partly reusable launch system. To reduce the cost of re-trieving booster, it is needed that the booster comes back to the launch site. This type of booster is often called a "fly-back booster". A fly-back booster has own wing, so it can fly to the launch site. Figure 7 shows an example of a fly-back booster concept for the Space Shuttle. In this book, RLV is defined as a system, in which at least the or-biter is reusable.

Figure 7: Fly-back Booster Concept (StarBooster, 2004)

1.2.3 Demand for RLVs

The International Space Station (ISS) has almost been completed and it is expected that normal operations will soon start on ISS. At present there are only two ways to transport humans to the ISS. One is by using the Space Shuttle of USA, and the other one is by using the Soyuz of Russia. Manned Space Transportation Systems require very high reliability at reasonable costs, but the present manned Space Transportation Systems do not have them enough to satisfy the safety and also cost requirement.

Some countries have studied or developed next generation Space Transportation Sys-tems including Reusable Launch Vehicles because RLVs are believed to be the key to reduce the launch cost. For example, NASA has studied many concepts like the X-series, and ESA has researched (Terrenoire, 1994) on the possibility of next generation booster system that flies back to launch site and can be used again after some mainte-nances.

A fully Reusable Launch Vehicle might provide the low cost access to space that is required to realize human exploration and colonization of the solar system. The Space Shuttle succeeded technically as a partially reusable launcher but failed economically. It requires several thousands support personal and takes at least two months to prepare for a launch. Shuttle launch costs are roughly $20 000 / kg, which is actually higher than most expendable launchers. There are many designs under development of robust RLV systems to bring the costs down to $2000 / kg on the short term and to $200 / kg on the longer term (Hobby Space, 2005).

1.2.4 Market Supply from Manufacturers

There are two trends in RLV development. One is developed by government like X-series of USA, and the other one is developed by private companies partly stimulated by organizations like X Prize. The X Prize is a $10 million prize to jumpstart the space tourism industry through competition between entrepreneurs and rocket experts in the world and was won by Scaled Composites' SpaceShipOne in 2004 as shown in Figure 8. The X Prize competition follows in the footsteps of more than 100 aviation incentive prizes offered between 1905 and 1935 that created today's multibillion-dollar air transport industry. (X Prize, 2005)

Figure 8: SpaceShipOne (X Prize, 2005)

Governments have been exploring the design and development of Reusable Launch Vehicles for several decades. For instance, NASA began the development of testbed vehicles such as the X-33, X-34 and X-37 that would prove technologies and operations concepts for next generation RLVs. The X-vehicles are designed to test technologies and operational concepts and therefore represent varied designs and mission profiles. Also several small "start-up" or "old economy" companies proposed to develop commercial RLVs such as Venture Star, Astroliner and K-1. Unfortunately, most of the development programs, independent if from government or private side, have been discontinued due to financial or technical problems.

Overall, there exist over 300 worldwide proposed vehicle concepts for RLVs, which could be realized by manufacturers from various countries summarized in Table 3. Derivative concepts with only slight changes are not listed. The majority of these vehicles are proposed for human space flight, but a few are unmanned. Table 4 shows 4 realized RLVs, namely X-15, Space Shuttle, Buran and SpaceShipOne. Buran has only been launched one time and therefore reusability has not been verified for this vehicle.

Table 3: Comparison of Worldwide RLV Concepts

Country	Suborbital	Orbital	Total
Argentina	1	0	1
Canada	2	1	3
China	0	3	3
France	1	14	15
Germany	3	18	21
India	0	1	1
Israel	1	0	1
Japan	1	8	9
Romania	1	0	1
Russia	3	50	53
United Kingdom	4	10	14
USA	46	132	178
Total	**63**	**237**	**300**

Table 4: Comparison of Worldwide RLV Realizations

Country	Suborbital	Orbital	Total
Russia	0	1 (Buran)	1
USA	2 (X-15, SpaceShipOne)	1 (Space Shuttle)	3
Total	**2**	**2**	**4**

1.2.5 Comparison of Reusable Launch Vehicles with Aircraft

Today's 24 years old Space Shuttle is a first generation, partially RLV. It is primarily used as an orbital scientific platform, for satellite deployment, retrieval and repair. Possible future vehicle characteristics are represented by the Hopper Plus concept and the Kankoh Maru Plus concept (Goehlich, 2003). The objectives of the Hopper Plus project are the reduction of operating and manufacturing costs and the enhancement of performance margins. It would primarily be used for space tourism flight and satellite deployment. Kankoh Maru Plus would introduce an era of space flight nearly as routine as today's air travel. It should enable new markets in Low Earth Orbit and provide multiple platforms for departure to new destinations. It is expected that for the successor generation of Kankoh Maru Plus there is nearly no distinction between a commercial airliner and a commercial launch vehicle and it would enable routine passenger space travel. Table 5 shows possible characteristic parameters of vehicle generations as forecasted by a study (Goehlich, 2003) compared with today's aircraft data.

Table 5: Possible Future Vehicle Characteristics

	Today	From 2015	From 2030	From 2060	Today
Vehicle Example	1. Generation: Space Shuttle	2. Generation: Hopper Plus	3. Generation: Kankoh Maru Plus	4. Generation: ?	B 747 Aircraft
Destination	LEO	Suborbit	LEO	LEO	Intercontinental
Launch Costs (average)	$20 000/kg	$2000/kg	$500/kg	$20/kg	$2/kg
Catastrophic Failure (max.)	1 in 100 flights	1 in 1000 flights	1 in 1000 flights	1 in 100 000 flights	1 in 2 000 000 flights
Passenger Escape	none	yes	yes	not required	not required
Fleet Flights per Year (max.)	6	100	2000	10 000	millions
Turnaround Time	5 months	1-2 weeks	2-10 days	6 hours	1 hour
Reusability	partial	full	full	full	full
Range Safety	flight unique	mission class unique	space traffic control	aerospace traffic control	air traffic control

1.2.6 Critical Comments about RLV Operations

It is assumed that RLVs are also used to transport space tourists to increase demand for flights (Goehlich, 2003) to be operated economically. With this precondition in mind, the following part discusses some critical points about RLVs.

Currently, aircraft show a failure rate of $0{,}5 \cdot 10^{-6}$ (1 in 2 000 000). Launch vehicles currently cannot yet match this as they have failure rates around 0,005 (1 in 200) at best. This is not favorable for space tourism, although there are many activities undertaken on ground with even greater risk factors. Due to the assumed high risk of space tourism ventures, insurance premiums might be very high. As long as insurance companies do not have coherent information about prices generated by implementation of space tourism, this will stay unchanged. Catastrophes like aviation disasters appear to carry more importance in people's minds than would be expected from general considerations of attitudes towards other forms of deaths (Moore, 1983). Even if the level of incident risk is low, the consequences when the risk occurs can be very large, which has implications for any insurance cover sought.

Tourists will demand greater comfort when traveling in space. Today's life support systems provide the necessary survival for astronauts but much is still to be desired to improve the habitability of spacecraft. Health concerns for short-term trips present no critical obstacle to space tourism. The most obvious lack of research is that of psychological stress on untrained space tourists.

Laws for public space activities must be established before space tourism can become a widespread activity. It is gradually becoming accepted that an appropriate regulatory framework for space passengers will be to treat it as an extension of aviation.

Although it is difficult to generalize about high-risk investment and its financing, some broad conclusion can be drawn. New companies are unable to raise risk finance through a stock exchange listing. There are specialized institutions whose main purpose is to provide venture or high-risk capital. Most, however, have maximum limits on the amount of capital they are prepared to provide in any single case. Since they are reluctant to invest in new company ventures, their services appeal mainly to existing companies (Moore, 1983).

1.3 Potential for International Collaboration

This section investigates on the pros and cons of an international Space Transportation System development, production and operation. Cultural business challenges of this international cooperation are discussed.

1.3.1 Motivation

In history, most rocket programs are national developments and only few are of international nature such as the Ariane family. The reason for this is that each country had the wish – due to political, military or economical reasons – to develop its own rocket program. Therefore, similar key technologies have been developed individually, each country with its own research budget but with similar objectives. Nowadays, the result is an oversupply of more or less similar national rockets on the one hand and a limited demand due to a stagnated satellite market on the other hand. Each country or company respectively is forced to reduce operation costs – painful by dismissing employees, reducing quality controls, etc. - to depreciate the big burden of development costs each nation has caused. To pass this process again for the development of a Reusable Launch Vehicle might not be wise. Instead, an international collaboration for only one Space Transportation System might be the alternative. Development costs can be shared between nations by only specializing in and being responsible for one segment of the rocket.

1.3.2 International Cooperation

In this section Arianespace is briefly introduced representing a typical international space organization. The main advantages and disadvantages caused by an international cooperation are pointed out.

Example: Arianespace

Commercial competition subsidized by governments has become an important part of space transportation competition. Europe, Russia, USA, China, Japan, India, etc. compete for the international space launch market. Each government has developed its own mechanisms for assisting its launch vehicles. For example, Arianespace is owned by 35 companies, 13 banks and the French national space agency CNES (Arianespace, 2005). The percentage distribution of Arianespace's shareholders in each country is shown in Figure 9. Although it operates as a private firm, Arianespace receives considerable indirect support from the European Space Agency, which has developed the various Ariane launchers, built the launch complexes and purchases launch services.

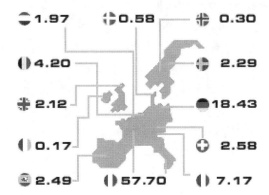

Figure 9: Arianespace's Shareholder Distribution in % (Arianespace, 2005)

Advantages

- Synergy effect: To achieve much more in a space program than a country can afford to attempt on its own for a given budget. It includes man power (experts from different countries), know-how (high technology) and use of existing infrastructures (spaceport, production facilities, etc.).

- Social benefit: In terms that it does broaden the mind and give people a planetary conscience (humankind is sitting in the same boat if e.g. a killer asteroid hits Earth).

- Realization: Because this might be the only way in near-term to put the vision into action.

Disadvantages

- Share pride with other nations: Because development of Space Transportation Systems is a national achievement that signals a nation's status as a space power, able to develop and use advanced technology, which is best reflected by an U.S. statement (Longsdon, 1989): "The space program is a visible symbol of U.S. world leadership; its challenges and accomplishments motivate scientific and technical excellence among U.S. students; and it provides for a diverse American population a sense of common national accomplishment and shared pride in American achievement."

- Share technology leadership: Is problematic because most launch technology has direct military applications and much of the technology has been classified or is sensitive.

- Tendency of parallel contractors: For an international cooperation, as well as for any other national program, it is necessary to have a clear-cut prime contractor / subcontractor relationship with well defined responsibilities. The lack of a strong prime contractor increase cost and probability for program failure. Cost growth is caused by more manpower, more interfaces, planned and unplanned parallel activities, schedule delays, etc. For example cost increase by 15 % if there are two parallel contractors instead of one, by 25 % for three parallel contractors, by 32 % for four parallel contractors, by 38 % for five parallel contractors and so on (Koelle, D.E., 2003). Due to political or prestige reason, this cost penalty is often accepted.

- Running board effect: This is existing in particular for international cooperation. It is difficult to monitor effort of another country and "force" a country to keep to a contract.

1.3.3 Cultural Business Challenges

In particular for an international cooperation there is a cultural business challenge beside the technical one. Management practices and effectiveness depend on cultural variables such as attitudes, beliefs, value systems, need hierarchies, etc, which are the result of different environmental factors in different countries. The challenge (which may also be transformed to a positive synergy) for managers and workers arises when these different environmental factors collide. In the following are given some examples for behavior based on different environmental factors. North American, European and Asian business cultures are only observed because this includes the leading countries in space technology and political power.

Management

Employees in high power distance cultures like Japan and China expect managers to lead and are less comfortable with the delegation of discretionary decisions than those from low power distance cultures such as USA. In addition, American or European managers take personal responsibility for their decisions (Ireland, 1991).

Decision-making

Western companies' decision-making is an individual process, while in Japanese companies it is a group process. Since many people are involved in the process and various meetings are held, there is a greater participation resulting in an easier and more efficient implementation. However, often too many meetings are held with many unnecessary questions and suggestions raised. These may significantly delay business decisions, which often require a swift response.

Family business

In the Chinese family business system, subordinates are supposed to think what the boss is thinking and tailor their ideas accordingly. Dissenting opinions and proposals are conveyed to the boss through personal channels with a duly respectful tone (Chen, 2004). But also in Western companies, this process for decision-making is often the case instead of objective assessment of cost and benefits for the company.

Success

While Americans see success as contingent upon their own individual efforts, a promotion in Japan as a reward for hard work may be detrimental to the employee's performance, as the highly cherished harmony between the promoted person and colleagues may be disturbed. Summarized, to be successful can be assumed to have the same meaning for the Asian and Western world but different for individuals of these two regions: While in the Western world success is mostly projected to the person itself, success within Asia is more related to the organization of the individual.

Alternatives

As for constructing alternatives, future-orientated cultures (typically Western cultures) tend to create more new options, whereas past-oriented cultures (typically Asian cultures) often search for a historical precedent.

Guanxi network

Representatives of other countries need acquire a basic understanding of guanxi dynamics for a successful cooperation with China, Japan and some other Asian countries. Guanxi seems to be the most important to understand business dealings in Asia. It can be best translated as friendship with implications of a continual exchange of favors. It is important to understand the difference between "guanxi network" in Asian countries and "protectionism" in western countries, which at first sight looks similar. In Asia, for thousands of years, it has been drummed into people that relationships, especially those within the family, are very important and the individual is less important. Children should learn to restrain themselves, to overcome their individuality so as to maintain harmony in the family. Therefore, guanxi network starts from the birth. Contrary, in western countries, the baby is born already with a strong ego, which is supported by the family with a result of an individual person with weak relationship to the family. Only launch government satellites with national launchers due to security and economic reasons is a typical example for protectionism. To sum it up it can be said that guanxi network is a tradition already there, while protectionism is a strategy gets through by government.

Fast-in and fast-out

The Chinese mentality "fast-in and fast-out" means the tendency to trade a smaller margin for a shorter sleeping period leads to a fast turnover orientation. Due to political uncertainties in China, the sleeping period is the most dangerous, as cash may never awake from its deep slumber. If the business deal requires a long slumber period, the Chinese businessman would demand much higher rates of return to justify his risk (Chen, 2004). Space programs have typically a long break-even point. An international cooperation contract with Chinese industry would mean that it might be overpaid due to the higher rates demanded for the reason given above.

Cost estimation

The cost from manufacturing to management in Japanese companies is already estimated at the stage of planning and design. The price that a customer is willing to pay for a product is first estimated and serves as the basis for calculating the prices of other component parts, ranging from designing to sale (Chen, 2004). In contrast, the typical method in USA and Europe is to design first, and then estimate cost based on a series of standard costs such as labor cost, material cost and manufacturing cost. Each item is calculated and is then put together by the accountant. If the cost is too high, the design will be modified and calculated again (Goehlich, 2003).

Motivation

In terms of motivation, Japanese and Korean employees seem to put more emphasis on extrinsic factors such as job security, work conditions and wages than on intrinsic factors like creativity and achievement. Japanese and Koreans are good at informal communication, but tend to be reluctant in expressing their views openly on formal occasions, especially when their opinions conflict with those of their superiors or colleagues. They are reluctant to convey bad news in a direct manner and carefully avoid open interpersonal conflicts. In addition, they pay careful attention to develop informal channels for communication (Chen, 2004). For an international cooperation, this may cause conflicts, because in order to avoid openly disagreement, Japanese tend to avoid discussions with the result that schedule will be delayed.

1.3.4 Conclusion

If the total costs for a reusable space transportation system or mission exceed a certain amount, an international cooperation may be the only alternative. An international cooperation has the advantage of sharing the total program costs. In practice, total costs in the case for an international program is higher compared to total costs in the case for a national program. This is caused by more manpower, more interfaces, planned and unplanned parallel activities, schedule delays, etc.

One major reason for these negative factors are the different strategies in each country of doing successful business. It can be considered that strategy is developed to achieve a fit between the organization and its environments (Chandler, 1979). Environmental factors are traditions, religion (e.g. Christianity, Confucianism, Buddhism), political conditions (e.g. safety, parties, regulations, ethnic conflicts), economic conditions (e.g. growth, recession, stagnation, productivity, income distribution, levels of income, employment rates, inflation, changes in market structure), social conditions (e.g. consumer attitudes), market geographic location, technology, etc. This means, each country has unconscious developed optimal business strategies. A mixture of these country specific strategies results in conflicts and, therefore, a decrease in overall benefit.

Part 2

Spaceships' Databank

Argentina

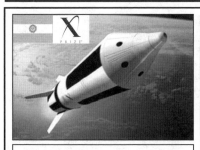

This suborbital vehicle with a capacity of 3 seats lands by using a parachute. The capsule is at least 12 times reusable.

Gauchito	★★☆☆☆
Developer	Pablo de Leon and Associates
Type	Suborbital
Number of Stages	1
Launch / Landing	Vertical / Vertical (Parachute)
Payload	1 crew + 2 pax (108 km)
Launch Mass	8 Mg
Propulsion System	4 x Rocket
Status	Active, 1984 - Present, Development
Image / Data Source	http://www.pablodeleon.com
E-mail	info@pablodeleon.com

Canada

Passengers will experience acceleration of no more than 4,5 g. This developer has its own training center.

Canadian Arrow	★★☆☆☆
Developer	Canadian Arrow
Type	Suborbital
Number of Stages	2 (RKT-e: 1,ORB-r: 1)
Launch / Landing	Vertical / Vertical (Parachute)
Payload	1 crew + 2 pax (112 km)
Launch Mass	14 Mg
Propulsion System	5 x Rocket
Status	Active, 2001 - Present, Development
Image / Data Source	http://www.canadianarrow.com
E-mail	info@canadianarrow.com

This is AVRO Canada's vehicle concept. After termination of this project Canadian engineers joined USA's manned flight program.

Space Threshold Vehicle	★★☆☆☆
Developer	AVRO Canada
Type	Orbital
Number of Stages	1
Launch / Landing	Horizontal / Horizontal
Payload	n.a.
Launch Mass	n.a.
Propulsion System	2 x Ramjet + 1 x Rocket
Status	Inactive, 1952 - n.a., Concept (Canceled)
Image / Data Source	http://www.cahs.com, R. L. Whitcomb
E-mail	n.a.

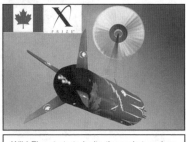

Wild Fire starts to ignite the rocket engines and ascent from a balloon at an altitude of 24 km.

Wild Fire	★★★☆☆
Developer	The da Vinci Project
Type	Suborbital
Number of Stages	2 (BLN-e: 1, ORB-r: 1)
Launch / Landing	Vertical / Vertical
Payload	2 pax (120 km)
Launch Mass	3,3 Mg
Propulsion System	2 x Rocket
Status	Active, 2000 - Present, Development
Image / Data Source	http://www.davinciproject.com
E-mail	bfeeney@davinciproject.com

China

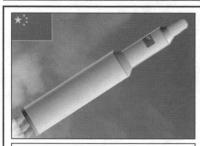

CZ RLV	★☆☆☆☆
Developer	n.a.
Type	Orbital
Number of Stages	2 (RKT-r: 1, ORB-r: 1)
Launch / Landing	Vertical / Vertical
Payload	2,0 Mg (200 km / 65 deg)
Launch Mass	400 Mg
Propulsion System	4 x Rocket
Status	Active, 2000 - n.a., n.a.
Image / Data Source	n.a.
E-mail	n.a.

The proposed rocket is a two-stage fully reusable launcher similar to Kistler's K-1. Its landing system includes a parachute and airbags.

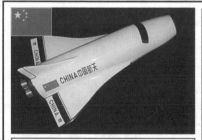

Project 921-3	★★★☆☆
Developer	n.a.
Type	Orbital
Number of Stages	2 (PLN-r: 1, ORB-r: 1)
Launch / Landing	Horizontal / Horizontal
Payload	2 crew + 6,0 Mg (500 km)
Launch Mass	330 Mg (Booster: 198 Mg, Orbiter: 132 Mg)
Propulsion System	Booster: 6 x Rocket + 8 x Ramjet, Orbiter: 4 x Rocket
Status	Inactive, 1992 - n.a., Development
Image / Data Source	http://www.astronautix.com
E-mail	eastronautica@hotmail.com

This is a Chinese space shuttle concept. This vehicle was planned to be operated in 2020.

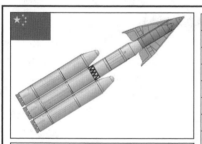

Tsien Spaceplane 1978 (Booster: CZ-2)	★☆☆☆☆
Developer	Tsien Hsue-shen
Type	Orbital
Number of Stages	3 (RKT-e: 2, ORB-r: 1)
Launch / Landing	Vertical / Horizontal
Payload	2 crew (185 km)
Launch Mass	500 Mg (CZ-2: 492 Mg, Orbiter: 7,8 Mg)
Propulsion System	CZ-2: Rocket, Orbiter: n.a.
Status	Inactive, 1978 - 1980, n.a. (Canceled)
Image / Data Source	http://www.astronautix.com
E-mail	eastronautica@hotmail.com

This vehicle concept strongly resembled the canceled US Dynasoar concept 15 years ago.

France

Aerospatiale VTVL RLV	★★★☆☆
Developer	Aerospatiale
Type	Orbital
Number of Stages	1
Launch / Landing	Vertical / Vertical
Payload	n.a.
Launch Mass	n.a.
Propulsion System	n.a.
Status	Inactive, 1990 - n.a., Concept
Image / Data Source	http://www.astronautix.com
E-mail	eastronautica@hotmail.com

n.a.

This concept is Aerospatiale`s VTVL SSTO study result.

Ares	★★★☆☆
Developer	Aerospatiale
Type	Orbital
Number of Stages	n.a. (RKT-e: n.a., ORB-r: 1)
Launch / Landing	Vertical / Horizontal
Payload	n.a.
Launch Mass	n.a. (Booster: n.a., Ares: 2 Mg)
Propulsion System	Booster: Rocket, Ares: n.a.
Status	Inactive, n.a. - n.a., n.a.
Image / Data Source	http://www.astronautix.com
E-mail	eastronautica@hotmail.com

This vehicle concept was merged with the Dassault Vehra to produce a single French proposal for the EXTV technology demonstrator.

EADS TSTO	★★★☆☆
Developer	EADS
Type	Orbital
Number of Stages	2 (FBB-r: 1, ORB-r: 1)
Launch / Landing	Vertical / Horizontal
Payload	n.a.
Launch Mass	n.a. (Booster: n.a., Orbiter: n.a.)
Propulsion System	Booster: n.a., Orbiter: n.a.
Status	Inactive, n.a. – n.a., Concept
Image / Data Source	http://www.eads.net
E-mail	n.a.

This vehicle is a French TSTO concept.

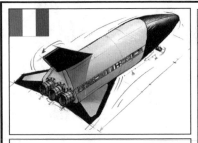

EXTV	★★★☆☆
Developer	ESA
Type	Suborbital
Number of Stages	1
Launch / Landing	Horizontal / Horizontal
Payload	n.a.
Launch Mass	4,2 Mg
Propulsion System	3 x Rocket
Status	Inactive, 1999 - n.a., Concept
Image / Data Source	http://www.esa.int/export/esaCP/index.html
E-mail	mailcom@esa.int

This vehicle means European eXperimental Test Vehicle. This is a rocket-powered re-entry demonstrator of the FESTIP study.

France

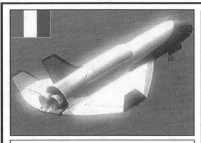

The FLTP (Future Launcher Technology Program) was an ESA program with responsibility assigned to CNES.

FLTP	★★★☆☆
Developer	ESA
Type	Orbital
Number of Stages	2 (FBB-r: 1, PLN-r: 1)
Launch / Landing	Vertical / Horizontal
Payload	n.a.
Launch Mass	n.a. (Booster: n.a., Orbiter: n.a.)
Propulsion System	Booster: n.a., Orbiter: n.a.
Status	Inactive, 1999 - n.a., n.a.
Image / Data Source	http://www.astronautix.com
E-mail	eastronautica@hotmail.com

France primarily led this program. But because of increasing estimated cost, this vehicle concept was canceled.

Hermes (Booster: Ariane 5)	★★☆☆☆
Developer	Dassault
Type	Orbital
Number of Stages	4 (RKT-e: 3, ORB-r: 1)
Launch / Landing	Vertical / Horizontal
Payload	3 crew + 3,0 Mg (800 km)
Launch Mass	737 Mg (Ariane 5: 714 Mg, Hermes: 23 Mg)
Propulsion System	Ariane 5: 4 x Rocket, Hermes: 2 x Rocket
Status	Inactive, 1985 - 1992, Development
Image / Data Source	http://www.dassault-aviation.com
E-mail	presse@dassault-aviation.fr

n.a.

Oriflamme is a French design for a scramjet-powered HTHL SSTO vehicle.

Oriflamme	★★☆☆☆
Developer	ESA
Type	Orbital
Number of Stages	1
Launch / Landing	Horizontal / Horizontal
Payload	n.a.
Launch Mass	n.a.
Propulsion System	1 x Scramjet
Status	Inactive, 1990 - n.a., Concept
Image / Data Source	http://www.astronautix.com
E-mail	eastronautica@hotmail.com

n.a.

The booster is powered by scramjets to Mach 12 before the second stage is separated.

Radiance	★★☆☆☆
Developer	ESA
Type	Orbital
Number of Stages	2 (PLN-r: 1, ORB-r: 1)
Launch / Landing	Horizontal / Horizontal
Payload	n.a.
Launch Mass	n.a. (Booster: n.a., Orbiter: n.a.)
Propulsion System	Booster: Scramjet, Orbiter: n.a.
Status	Inactive, 1990 - n.a., Concept
Image / Data Source	http://www.astronautix.com
E-mail	eastronautica@hotmail.com

France

At Mach 6 the Hermes spaceplane with an expendable second stage is separated from a hypersonic first stage.

STAR-H (Orbiter: Hermes)	★★☆☆☆
Developer	ESA
Type	Orbital
Number of Stages	3 (PLN-r: 1, RKT-e: 1, ORB-r: 1)
Launch / Landing	Horizontal / Horizontal
Payload	n.a.
Launch Mass	n.a. (Booster: n.a., Booster: n.a., Hermes: 23 Mg)
Propulsion System	Booster: n.a., Booster: n.a., Hermes: 2 x Rocket
Status	Inactive, 1990 - n.a., n.a.
Image / Data Source	http://www.astronautix.com
E-mail	eastronautica@hotmail.com

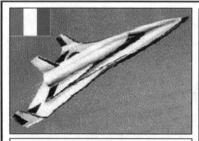

This system is a SSTO ramjet and rocket mix power HTHL study of the 1980s.

STS-2000 SSTO	★★☆☆☆
Developer	ESA
Type	Orbital
Number of Stages	1
Launch / Landing	Horizontal / Horizontal
Payload	n.a.
Launch Mass	n.a.
Propulsion System	Ramjet + Rocket
Status	Inactive, 1987 - n.a., n.a.
Image / Data Source	http://www.astronautix.com
E-mail	eastronautica@hotmail.com

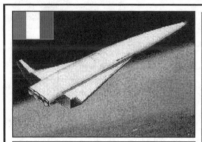

This system is a HTHL variant of STS-2000 and has a ramjet and rocket mixed power first stage. At mach 6 the orbiter will be separated from the first stage.

STS-2000 TSTO	★★★☆☆
Developer	ESA
Type	Orbital
Number of Stages	2 (PLN-r: 1, ORB-r: 1)
Launch / Landing	Horizontal / Horizontal
Payload	n.a.
Launch Mass	n.a.
Propulsion System	Booster: Ramjet + Rocket, Orbiter: Rocket
Status	Inactive, 1987 - n.a., n.a.
Image / Data Source	http://www.astronautix.com
E-mail	eastronautica@hotmail.com

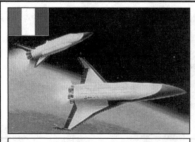

Taranis is a French study of a VTHL TSTO vehicle with an expendable orbiter fuel tank.

Taranis	★★★☆☆
Developer	n.a.
Type	Orbital
Number of Stages	2 (FBB-r: 1, ORB-r: 1)
Launch / Landing	Vertical / Horizontal
Payload	n.a.
Launch Mass	n.a. (Booster: n.a., Orbiter: n.a.)
Propulsion System	Booster: n.a., Orbiter: n.a.
Status	Inactive , 1990 - n.a., Concept
Image / Data Source	http://www.astronautix.com
E-mail	eastronautica@hotmail.com

France

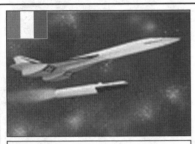

TAS is a concept of the 1960s.

TAS	★★☆☆☆
Developer	Dassault
Type	Orbital
Number of Stages	2 (PLN-r: 1, ORB-r: 1)
Launch / Landing	Horizontal / Horizontal
Payload	n.a.
Launch Mass	n.a. (Booster: n.a., Orbiter: n.a.)
Propulsion System	Booster: n.a., Orbiter: n.a.
Status	Inactive, 1960 - n.a., n.a.
Image / Data Source	http://www.dassault-aviation.com/espace
E-mail	philippe.coue@dassault-aviation.fr

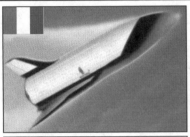

Themis was a planned ESA booster stage demonstrator to validate integrated propellant tank technology necessary for a reusable Ariane 5 successor.

Themis	★★★☆☆
Developer	ESA
Type	Orbital
Number of Stages	2 (RKT-e: 1. ORB-r: 1)
Launch / Landing	Vertical / Horizontal
Payload	n.a.
Launch Mass	n.a. (Booster: n.a., Themis: 55 Mg)
Propulsion System	Booster: n.a., Themis: n.a.
Status	Inactive, 1998 - n.a., n.a.
Image / Data Source	http://www.astronautix.com
E-mail	eastronautica@hotmail.com

To lay the groundwork for RLV, the company has studied the Vehra demonstrator family.

Vehra (Booster: Airbus A 300)	★★☆☆☆
Developer	Dassault
Type	Orbital
Number of Stages	2 (PLN-r: 1, ORB-r: 1)
Launch / Landing	Horizontal / Horizontal
Payload	0,3 Mg (185 km)
Launch Mass	165 Mg (Airbus A 300: 138 Mg, Vehra: 27 Mg)
Propulsion System	Airbus A 300: 2 x Turbojet, Vehra: Rocket
Status	Inactive, n.a. - n.a., n.a.
Image / Data Source	http://www.dassault-aviation.com/espace
E-mail	philippe.coue@dassault-aviation.fr

Germany

n.a.

Astros	★★☆☆☆
Developer	ESA
Type	Orbital
Number of Stages	1
Launch / Landing	Horizontal / Horizontal
Payload	n.a.
Launch Mass	n.a.
Propulsion System	n.a.
Status	Inactive, 1994 - 1999, n.a. (Canceled)
Image / Data Source	http://www.astronautix.com
E-mail	eastronautica@hotmail.com

Astros is a HTHL sled-launched SSTO vehicle concept designed under the Future European Space Transportation Investigation Program (FESTIP).

BETA	★☆☆☆☆
Developer	MBB
Type	Orbital
Number of Stages	1
Launch / Landing	Vertical / Vertical
Payload	1 crew + 2,0 Mg (90 km / 6 deg)
Launch Mass	130 Mg
Propulsion System	12 x Rocket
Status	Inactive, 1969 - n.a., n.a. (Canceled)
Image / Data Source	http://www.astronautix.com
E-mail	eastronautica@hotmail.com

In 1969 Dietrich Koelle sketched out a reusable VTVL design called BETA using Bono's SASSTO concept as a starting point.

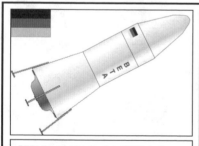

BETA II	★★★☆☆
Developer	MBB
Type	Orbital
Number of Stages	1
Launch / Landing	Vertical / Vertical
Payload	18,0 Mg (90 km / 6 deg)
Launch Mass	600 Mg
Propulsion System	Rocket
Status	Inactive, 1986 - n.a., n.a. (Concept)
Image / Data Source	http://www.astronautix.com
E-mail	eastronautica@hotmail.com

BETA II is Dietrich Koelle's SSTO design.

n.a.

BETA III	★★★☆☆
Developer	MBB
Type	Orbital
Number of Stages	1
Launch / Landing	Vertical / Vertical
Payload	5 crew + 100 pax or 17,0 Mg (90 km / 6 deg)
Launch Mass	780 Mg
Propulsion System	25 x Rocket
Status	Inactive, 1996 - 1998, n.a.
Image / Data Source	http://www.astronautix.com
E-mail	eastronautica@hotmail.com

From 1996 to 1998 Dietrich Koelle updated the design of BETA for using it as an ISS resupply vehicle in place of the shuttle and as a space tourism vehicle for 100 pax.

Germany

BETA IV ★★★☆☆

Developer	MBB
Type	Orbital
Number of Stages	1
Launch / Landing	Vertical / Vertical
Payload	100 Mg (90 km / 6 deg)
Launch Mass	2000 Mg
Propulsion System	Rocket
Status	Inactive, 1986 - n.a., n.a.
Image / Data Source	http://www.astronautix.com
E-mail	eastronautica@hotmail.com

n.a.

This system is Dietrich Koelle's largest SSTO concept. In 1986 he updated the design for use of a more sophisticated trajectory analysis.

C-50 ★★☆☆☆

Developer	EADS Space Transportation
Type	Orbital
Number of Stages	n.a.
Launch / Landing	n.a.
Payload	42 pax
Launch Mass	n.a.
Propulsion System	n.a.
Status	Active, n.a. - Present, n.a.
Image / Data Source	http://www.eads.net
E-mail	contact@astrium.eads.net

n.a.

This rocket concept is designed to transport 42 space tourists.

DSL ★★☆☆☆

Developer	DLR
Type	Orbital
Number of Stages	2 (PLN-r: 1, ORB-r: 1)
Launch / Landing	Horizontal / Horizontal
Payload	16,1 Mg (400 km / 29 deg)
Launch Mass	560 Mg (Booster: n.a., Orbiter: n.a.)
Propulsion System	Booster: 10 x Turbofan, Orbiter: Rocket
Status	Active, 1992 - Present, n.a.
Image / Data Source	http://www.kp.dlr.de/DSL/Welcome.html
E-mail	dsl@dlr.de

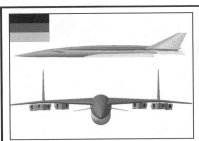

The orbiter will be separated from the supersonic carrier aircraft. This system has various second stage options.

EARL 1 ★★☆☆☆

Developer	Dornier
Type	Orbital
Number of Stages	2 (RKT-r: 1, ORB-r: 1)
Launch / Landing	Vertical / Horizontal
Payload	5,4 Mg (LEO)
Launch Mass	270 Mg (Booster: n.a., Orbiter: n.a.)
Propulsion System	Booster: Rocket, Orbiter: n.a.
Status	Inactive, 1987 - 1990, n.a.
Image / Data Source	http://www.astronautix.com
E-mail	eastronautica@hotmail.com

The orbiter is mounted on top of the booster.

Germany

This system is a later EARL version from 1990 with parallel staging. Both stages were winged and recoverable.

EARL 2	★★★☆☆
Developer	ESA
Type	Orbital
Number of Stages	2 (FBB-r: 1, ORB-r: 1)
Launch / Landing	n.a. / Horizontal
Payload	n.a.
Launch Mass	n.a. (Booster: n.a., Orbiter: n.a.)
Propulsion System	Booster: n.a., Orbiter: n.a.
Status	Inactive, 1990 - n.a., n.a.
Image / Data Source	http://www.astronautix.com
E-mail	eastronautica@hotmail.com

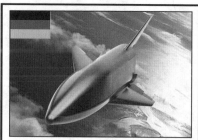

This vehicle program is not an official ESA program. It is planned to start operation from 2015.

Hopper (suborbital)	★★★★☆
Developer	EADS Space Transportation
Type	Suborbital
Number of Stages	1
Launch / Landing	Horizontal / Horizontal
Payload	7,5 Mg
Launch Mass	491 Mg
Propulsion System	3 x Rocket
Status	Active, n.a. - Present, Concept
Image / Data Source	http://www.flug-revue.rotor.com
E-mail	vthomalla@motorpresse.de

Hopper Plus is a RLV concept for space tourism. It can transport 30 passengers.

Hopper Plus	★★★★☆
Developer	EADS Space Transportation, Robert A. Goehlich
Type	Suborbital
Number of Stages	1
Launch / Landing	Horizontal / Horizontal
Payload	30 pax
Weight	460 Mg
Propulsion System	3 x Rocket
Status	Active, n.a. - Present, Concept
Image / Data Source	http://www.robert-goehlich.de
E-mail	mail@robert-goehlich.de

Following the cancelation of Saenger II, Germany considered a manned X-15 type flight test vehicle called "Hytex" capable of Mach 6 flight.

Hytex	★★☆☆☆
Developer	MBB
Type	Suborbital
Number of Stages	1
Launch / Landing	Horizontal / Horizontal
Payload	n.a.
Launch Mass	n.a.
Propulsion System	n.a.
Status	Inactive, 1995 - n.a., n.a. (Canceled)
Image / Data Source	http://www.astronautix.com
E-mail	eastronautica@hotmail.com

Germany

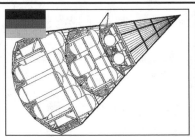

LART	★★☆☆☆
Developer	MBB
Type	Orbital
Number of Stages	1
Launch / Landing	Horizontal / Horizontal
Payload	n.a.
Launch Mass	n.a.
Propulsion System	Jet + Rocket
Status	Inactive, 1985 - n.a., n.a.
Image / Data Source	http://www.astronautix.com
E-mail	eastronautica@hotmail.com

This system is MBB`s and ERNO`s airbreathing HTHL SSTO rocket proposal from the mid-1980s and is largely similar to the BAe HOTOL.

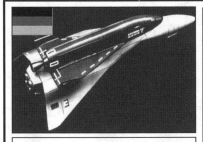

NEPTUNE	★★★☆☆
Developer	TU Berlin, Heinz-Hermann Koelle
Type	Orbital
Number of Stages	3 (RKT-e: 2, ORB-r: 1)
Launch / Landing	n.a.
Payload	350,0 Mg (LEO)
Launch Mass	6000 Mg
Propulsion System	57 x SSME Rocket
Status	Active, n.a. - Present, n.a.
Image / Data Source	http://www.tu-berlin.de
E-mail	101612.2141@compuserve.com

The NEPTUNE vehicle is a typical Heavy Lift Launch vehicle (HLLV). The developer thought it as a key element of the space transportation of the next century.

Saenger I	★★★☆☆
Developer	Saenger
Type	Orbital
Number of Stages	2 (PLN-r: 1, ORB-r: 1)
Launch / Landing	Horizontal / Horizontal
Payload	n.a.
Launch Mass	n.a. (Booster: n.a., Orbiter: n.a.)
Propulsion System	Booster: n.a., Orbiter: n.a.
Status	Inactive, 1964 - n.a., n.a.
Image / Data Source	http://www.astronautix.com
E-mail	eastronautica@hotmail.com

This system is a German winged HTHL TSTO launch vehicle.

Saenger II (Orbiter: Horus)	★★★☆☆
Developer	MBB
Type	Orbital
Number of Stages	2 (PLN-r: 1, ORB-r: 1)
Launch / Landing	Horizontal / Horizontal
Payload	2 crew + 3,0 Mg (LEO)
Launch Mass	340 Mg (Booster: n.a., Horus: n.a.)
Propulsion System	Booster: 6 x Turbo-ramjet, Horus: 1 x Rocket
Status	Inactive, 1985 - 1991, Development (Canceled)
Image / Data Source	http://www.astronautix.com
E-mail	eastronautica@hotmail.com

This system represents an advanced concept that is not feasible with today's technology. It has been studied from 1985 to 1991 by MBB.

Germany

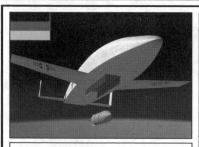

Silver Bird	★★☆☆☆
Developer	Eugen Saenger
Type	Orbital
Number of Stages	n.a.
Launch / Landing	n.a.
Payload	1,0 Mg
Launch Mass	100 Mg
Propulsion System	n.a.
Status	Inactive, 1944 - n.a., n.a.
Image / Data Source	n.a.
E-mail	n.a.

Silver Bird can be considered as the first RLV concept in history.

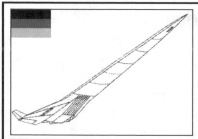

Vector	★☆☆☆☆
Developer	TU Berlin
Type	Orbital
Number of Stages	n.a.
Launch / Landing	n.a. / Horizontal
Payload	120 pax
Launch Mass	385 Mg
Propulsion System	n.a.
Status	Inactive, 1995 - n.a., n.a.
Image / Data Source	n.a.
E-mail	n.a.

This RLV concept has advanced airbreathing engines and is capable to transport 120 passengers.

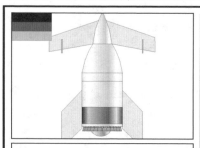

Von Braun RLV 1948	★☆☆☆☆
Developer	Von Braun
Type	Orbital
Number of Stages	3 (RKT-e: 2, ORB-r: 1)
Launch / Landing	Vertical / Horizontal
Payload	25,0 Mg (1730 km / 24 deg)
Launch Mass	6400 Mg (Booster: 6270 Mg, Orbiter: 130 Mg)
Propulsion System	Booster: 78 x Rocket, Orbiter: 1 x Rocket
Status	Inactive, 1948 - n.a., Concept
Image / Data Source	http://www.astronautix.com
E-mail	eastronautica@hotmail.com

This concept is Von Braun's 1948 design for a reusable space launcher. This was partly driven by the need for large parachute canisters.

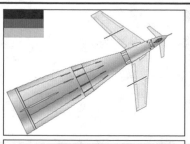

Von Braun RLV 1952	★☆☆☆☆
Developer	Von Braun
Type	Orbital
Number of Stages	3 (RKT-e: 2, ORB-r: 1)
Launch / Landing	Vertical / Horizontal
Payload	25,0 Mg (1730 km / 24 deg)
Launch Mass	6400 Mg (Booster: 6270 Mg, Orbiter: 130 Mg)
Propulsion System	Booster: 85 x Rocket, Orbiter: 5 x Rocket
Status	Inactive, 1952 - n.a., Concept
Image / Data Source	http://www.astronautix.com
E-mail	eastronautica@hotmail.com

This concept is Von Braun's 1952 design using the same mass and performance calculations done in 1948.

Germany

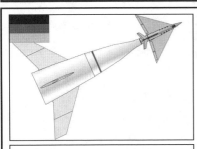

In 1956 for the book "Exploration of Mars" the 1952 design was significantly down-sized. The first and second stage is reduced to 20 % of their former size.

Von Braun RLV 1956	★☆☆☆☆
Developer	Von Braun
Type	Orbital
Number of Stages	3 (RKT-e: 2, ORB-r: 1)
Launch / Landing	Vertical / Horizontal
Payload	13,6 Mg (102 km / 24 deg)
Launch Mass	1380 Mg (Booster: 1354 Mg, Orbiter: 26 Mg)
Propulsion System	Booster: 37 x Rocket, Orbiter: 1 x Rocket
Status	Inactive, 1956 - n.a., Concept
Image / Data Source	http://www.astronautix.com
E-mail	eastronautica@hotmail.com

India

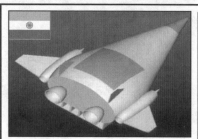

AVATAR will support Space Solar Power (SSP) stations affordable providing a global solution for the coming energy crisis.

AVATAR	★★☆☆☆
Developer	DRDO
Type	Orbital
Number of Stages	1
Launch / Landing	Horizontal / Horizontal
Payload	1,0 Mg (100 km)
Launch Mass	45 Mg
Propulsion System	1 x Turbo-ramjet + 1 x Scramjet + 1 x Rocket
Status	Active, 1987 - Present, Concept
Image / Data Source	http://www.geocities.com/spacetransport
E-mail	spacetransport@yahoo.com

Israel

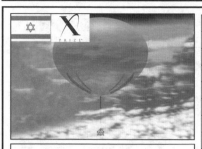

The NEGEV-5 will be a self-sufficient reusable vehicle capable of being launched and recovered anywhere in the world from land or sea without runway or assistant.

NEGEV-5	★★★☆☆
Developer	IL Aerospace Technologies
Type	Suborbital
Number of Stages	2 (BLN-e: 1, ORB-r: 1)
Launch / Landing	Vertical / Vertical (Parachute)
Payload	1 crew + 2 pax or 0,7 Mg (120 km)
Launch Mass	3,4 Mg
Propulsion System	1 x Rocket
Status	Active, 2002 - Present, Development
Image / Data Source	http://www.ilat.net
E-mail	info@ilat.co.il

Japan

The HOPE (H-2 Orbiting Plane) is a concept of a Japanese reusable unmanned winged spacecraft. Some technologies that are needed to realize HOPE were done.

HOPE (Booster: H-2)	★★★★☆
Developer	JAXA
Type	Orbital
Number of Stages	4 (RKT-e: 3, ORB-r: 1)
Launch / Landing	Vertical / Horizontal
Payload	2,0 Mg (LEO)
Launch Mass	422 Mg (H-2: 410 Mg, HOPE: 12 Mg)
Propulsion System	H-2: 4 x Rocket, HOPE: Rocket
Status	Inactive, 1986 - 2002, Development (Canceled)
Image / Data Source	http://www.astronautix.com/craft/hope.htm
E-mail	proffice@jaxa.jp

JAXA is studying the HOPE-X for the purpose to demonstrate state-of-the-art space transportation technologies.

HOPE-X (Booster: H-2A)	★★☆☆☆
Developer	JAXA
Type	Orbital
Number of Stages	4 (RKT-e: 3, ORB-r: 1)
Launch / Landing	Vertical / Horizontal
Payload	1,0 Mg (200 km)
Launch Mass	265 Mg (H-2A: 260 Mg, HOPE-X: 5 Mg)
Propulsion System	H-2A : 4 x Rocket, HOPE-X : 2 x Rocket
Status	Active, 1993 - Present, Development
Image / Data Source	http://www.jaxa.jp/index_e.html
E-mail	proffice@jaxa.jp

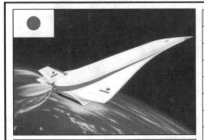

Mitsubishi Heavy Industry proposed this space plane to NASDA that has been the former JAXA.

Japanese Space Plane	★★★☆☆
Developer	JAXA, Mitsubishi Heavy Industry
Type	Orbital
Number of Stages	1
Launch / Landing	Horizontal / Horizontal
Payload	10 pax + 20,0 Mg (200 km)
Launch Mass	350 Mg
Propulsion System	1 x Scramjet + 1 x Rocket
Status	Active, 1995 - Present, Concept
Image / Data Source	http://www.jaxa.jp/index_e.html
E-mail	proffice@jaxa.jp

This concept is designed for X Prize organized by students.

JEDI Project	★★★★☆
Developer	Shibaura Institute of Technology
Type	Suborbital
Number of Stages	2 (PLN-r: 1, ORB-r: 1)
Launch / Landing	Horizontal / Horizontal
Payload	n.a.
Launch Mass	n.a. (Booster: n.a., Orbiter: n.a.)
Propulsion System	Booster: n.a., Orbiter: n.a.
Status	Inactive, n.a. - n.a., Concept
Image / Data Source	http://www.shibaura-it.ac.jp
E-mail	webinfo@ow.shibaura-it.ac.jp

Japan

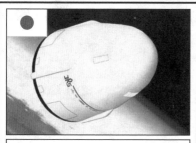

Kawasaki designed for SSTO a reusable vehicle. This would carry 50 passengers to orbiting hotels or fast intercontinental orbital flight.

Kankoh Maru	★★★★★
Developer	Kawasaki Heavy Industries
Type	Orbital
Number of Stages	1
Launch / Landing	Vertical / Vertical
Payload	4 crew + 50 pax (200 km)
Launch Mass	550 Mg
Propulsion System	12 x Rocket
Status	Inactive, 1993 - n.a., Concept (Canceled)
Image / Data Source	http://www.astronautix.com
E-mail	eastronautica@hotmail.com

JAXA is developing next generation reusable launch vehicles. This concept is one of them. The 1st stage has high performance jet engines named "ATREX".

NAL TSTO Spaceplane	★★☆☆☆
Developer	JAXA
Type	Orbital
Number of Stages	2 (PLN-r: 1, ORB-r: 1)
Launch / Landing	Horizontal / Horizontal
Payload	8 Mg (LEO)
Launch Mass	400 Mg (Booster: 296 Mg, Orbiter: 104 Mg)
Propulsion System	Booster: Airbreathing Engine, Orbiter: Rocket
Status	Active, n.a. - Present, Concept
Image / Data Source	http://www.nal.go.jp/strpc/jpn/index_h.html
E-mail	proffice@jaxa.jp

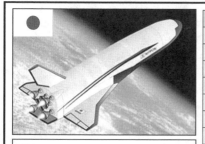

This concept is a rocket only propulsion spaceplane for the future space transportation.

Rocket Plane a	★★★☆☆
Developer	JAXA
Type	Orbital
Number of Stages	1
Launch / Landing	Horizontal / Horizontal
Payload	n.a.
Launch Mass	n.a.
Propulsion System	Rocket
Status	Active, n.a. - Present, Concept
Image / Data Source	http://www.jaxa.jp
E-mail	proffice@jaxa.jp

This vehicle has a unique wing called WSDT. WSDT is an expendable Wing Shaped Drop Tank. In the way to the space, this vehicle drops the WSDT.

Rocket Plane b (Booster: Ekranoplane)	★★★★☆
Developer	Keio University, Yoshiaki Ohkami
Type	Orbital
Number of Stages	2 (PLN-r: 1, ORB-r: 1)
Launch / Landing	Horizontal / Horizontal
Payload	3 crew (250 km)
Launch Mass	n.a. (Ekranoplane: n.a., Orbiter: 250 Mg)
Propulsion System	Ekranoplane: Airbreathing Engine, Orbiter: 3 x Rocket
Status	Active, 2002 - Present, Development
Image / Data Source	http://www.ohkami.sd.keio.ac.jp
E-mail	ohkami@sd.keio.ac.jp

Japan

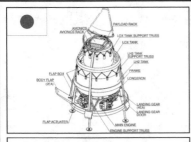

JAXA's proposal is a fully reusable VTVL sounding rocket. An experimental demonstrator vehicle called "RVT" went through two hardware experiments.

RSR - Reusable Sounding Rocket ★★★★★	
Developer	JAXA
Type	Orbital
Number of Stages	1
Launch / Landing	Vertical / Vertical
Payload	0,1 Mg (300 km)
Launch Mass	3,8 Mg
Propulsion System	4 x Rocket
Status	Active, n.a. - Present, Development
Image / Data Source	http://www.jaxa.jp
E-mail	proffice@jaxa.jp

Romania

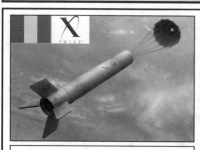

The Orizont vehicle is formed of a rocket and a capsule.

Orizont	★★☆☆☆
Developer	ARCA
Type	Suborbital
Number of Stages	1
Launch / Landing	Vertical / Vertical (Parachute)
Payload	1 crew + 2 pax (100 km)
Launch Mass	7,0 Mg
Propulsion System	1 x Rocket
Status	Active, 1998 - Present, Development
Image / Data Source	http://www.arcaspace.ro/en/home.htm
E-mail	office@arcaspace.ro

Russia

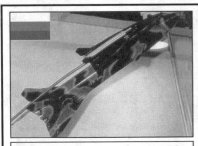

This system is a small HTHL TSTO system proposed for the Russian Air Force in 1993.

17K-AM	★★☆☆☆
Developer	Chelomei
Type	Orbital
Number of Stages	2 (PLN-r: 1, ORB-r: 1)
Launch / Landing	Horizontal / Horizontal
Payload	n.a.
Launch Mass	n.a. (Booster: n.a., Orbiter: n.a.)
Propulsion System	Booster: n.a., Orbiter: n.a.
Status	Inactive, 1993 - n.a., n.a.
Image / Data Source	http://www.astronautix.com
E-mail	eastronautica@hotmail.com

This system is a sled-launched, air-breathing HTHL SSTO vehicle proposed in Russia.

Ajax	★☆☆☆☆
Developer	Molniya
Type	Orbital
Number of Stages	1
Launch / Landing	Horizontal / Horizontal
Payload	n.a.
Launch Mass	n.a.
Propulsion System	n.a.
Status	Inactive, 1993 - n.a., n.a.
Image / Data Source	http://www.astronautix.com
E-mail	eastronautica@hotmail.com

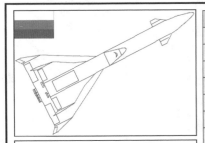

This system is a hydrofoil-launched, winged recoverable first and second stage.

Albatros	★★☆☆☆
Developer	Alekseyev, Sukhoi
Type	Orbital
Number of Stages	2 (PLN-r: 1, ORB-r: 1)
Launch / Landing	Horizontal / Horizontal
Payload	30 Mg (220 km)
Launch Mass	2330 Mg (Booster: 2010 Mg, Orbiter: 320 Mg)
Propulsion System	Booster: 4 x Rocket, Orbiter: 4 x Rocket
Status	Inactive, 1974 - n.a., Concept
Image / Data Source	http://www.astronautix.com
E-mail	eastronautica@hotmail.com

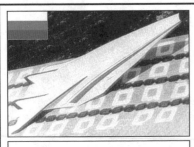

This system is a sled-launched airbreathing HTHL SSTO launch vehicle proposed in Russia.

ASA	★☆☆☆☆
Developer	n.a.
Type	Orbital
Number of Stages	1
Launch / Landing	Horizontal / Horizontal
Payload	n.a.
Launch Mass	n.a.
Propulsion System	n.a.
Status	Inactive, 1993 - n.a., Concept
Image / Data Source	http://www.astronautix.com
E-mail	eastronautica@hotmail.com

Russia

Atlant System is based on Atlant VM-T technologies. It is similar to Myasishchev RLV. The difference of them is Atlant's orbiter.

Atlant System (Booster: Atlant VM-T)	★★☆☆☆
Developer	Myasishchev Design Bureau
Type	Orbital
Number of Stages	3 (PLN-r: 2, ORB-r: 1)
Launch / Landing	Horizontal / Horizontal
Payload	1,3 Mg
Launch Mass	182 Mg (Atlant VM-T: n.a., Booster: n.a., Orbiter: n.a.)
Propulsion System	Atlant VM-T: 4 x Turbofan, Booster: n.a., Orbiter: Rocket
Status	Active, n.a. - Present, n.a.
Image / Data Source	http://www.corbina.net/~kluka/Emz/Emz0-e.htm
E-mail	mdb@mail.sitek.ru

Bizan is a Soviet air-launched spaceplane design iteration between the "49" and "MAKS" concepts.

Bizan (Booster: An-124)	★★★☆☆
Developer	Molniya
Type	Orbital
Number of Stages	3 (PLN-r: 1, RKT-e: 1, ORB-r: 1)
Launch / Landing	Horizontal / Horizontal
Payload	1 crew + 4,0 Mg (1000 km / 94 deg)
Launch Mass	710 Mg (An-124: 510 Mg, Booster: 185 Mg, Bizan: 15 Mg)
Propulsion System	An-124: 4 x Turbofan, Booster: 2 x Rocket, Bizan: 1 x Rocket
Status	Inactive, 1981 - n.a., n.a.
Image / Data Source	http://www.astronautix.com
E-mail	eastronautica@hotmail.com

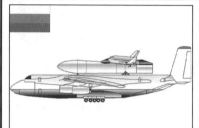

This system is a scale model of Spiral OS to investigate on the hypersonic aerodynamic characteristics and heat shield materials of the manned Spiral OS lifting body.

BOR-4	★★☆☆☆
Developer	Molniya
Type	Orbital
Number of Stages	n.a.
Launch / Landing	n.a. / Horizontal
Payload	0 Mg
Launch Mass	n.a. (Booster: n.a., BOR-4: 1,2 Mg)
Propulsion System	n.a.
Status	Inactive, 1973 - 1984, Realized (4 flights)
Image / Data Source	http://www.astronautix.com
E-mail	eastronautica@hotmail.com

BOR-5 was made for investigating the aerodynamic characteristics of Buran at hypersonic speeds and is a 1:8 sub-scale model of Buran.

BOR-5 (Booster: Kosmos 65MP)	★★☆☆☆
Developer	Molniya
Type	Suborbital
Number of Stages	3 (RKT-e: 2, ORB-r: 1)
Launch / Landing	Vertical / Horizontal
Payload	0 Mg
Launch Mass	109 Mg (Kosmos 65MP: 108 Mg, BOR-5: 1,4 Mg)
Propulsion System	Kosmos 65MP: 2 x Rocket, BOR-5: n.a.
Status	Inactive, 1983 - 1988, Realized (6 flights)
Image / Data Source	http://www.astronautix.com
E-mail	eastronautica@hotmail.com

Russia

This system is very similar to the Space Shuttle System. It was canceled in 1991 due to political reasons.

Buran (Booster: Energia)	★★★★☆
Developer	Defence Ministry, Korolev
Type	Orbital
Number of Stages	2 (RKT-e: 1, ORB-r: 1)
Launch / Landing	Vertical / Horizontal
Payload	30,0 Mg (1000 km / 94 deg)
Launch Mass	2525 Mg (Energia: 2420 Mg, Orbiter: 105 Mg)
Propulsion System	Energia: 5 x Rocket, Orbiter: 2 x Rocket
Status	Inactive, 1976 - 1993, Realized (1 Flight)
Image / Data Source	http://www.buran.ru/htm/molniya.htm
E-mail	molniya@dol.ru

This system is a fully recoverable version of the Energia launch vehicle with four winged boosters and a winged core stage.

Buran-T	★★★☆☆
Developer	Korolev
Type	Orbital
Number of Stages	2 (FBB-r: 1, ORB-r: 1)
Launch / Landing	Vertical / Horizontal
Payload	n.a.
Launch Mass	n.a. (Booster: n.a., Orbiter: n.a.)
Propulsion System	Booster: n.a., Orbiter: n.a.
Status	Inactive, 1989 - n.a., n.a.
Image / Data Source	http://www.energia.ru/english
E-mail	mail@rsce.ru

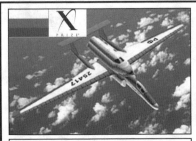

C-21 will be separated from M-55X at the altitude of 17 km. Suborbital Corporation estimates it will cost $10 million to build and test the system.

Cosmopolis XXI (Booster: M-55X, Orbiter: C-21)	★★★★★
Developer	Suborbital Corporation
Type	Suborbital
Number of Stages	2 (PLN-r: 1, ORB-r: 1)
Launch / Landing	Horizontal / Horizontal
Payload	2 crew + 2 pax (100 km)
Launch Mass	29 Mg (M-55X: 27 Mg, C-21: 2 Mg)
Propulsion System	M-55X: 2 x Turbojet, C-21: 1 x Rocket
Status	Active, n.a. - Present, Development
Image / Data Source	http://www.spaceadventure.com
E-mail	info@spaceadventures.com

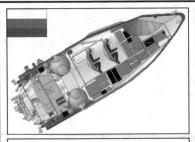

The Kliper manned spacecraft replacement for Soyuz has a 14,5 Mg reusable lifting body as a space station ferry and lifeboat. Kliper is planned to be operated from 2010.

Kliper (Booster: Onega, Orbiter: Kliper)	★★★★★
Developer	RKK Energia
Type	Orbital
Number of Stages	3 (RKT-e: 2, ORB-r: 1)
Launch / Landing	Vertical / Horizontal
Payload	2 crew + 4 pax + 0,7 Mg
Launch Mass	n.a. (Onega: n.a., Kliper: 15 Mg)
Propulsion System	Onega: Rocket, Kliper: n.a.
Status	Active, 2004 - Present, Concept
Image / Data Source	http://www.astronautix.com
E-mail	eastronautica@hotmail.com

Russia

LII - the Gromov Experimental Flight Institute at Zhukoskiy - designed several alternate spaceplane concepts for air-launch from the Atlant VM-T.

LII Spaceplane (Booster: Atlant VM-T) ★★★☆☆	
Developer	LII
Type	Orbital
Number of Stages	3 (PLN-r: 2, ORB-r: 1)
Launch / Landing	Horizontal / Horizontal
Payload	1 crew
Launch Mass	600 Mg (Atlant VM-T: n.a., Booster: n.a., Orbiter: n.a.)
Propulsion System	Atlant VM-T: 4 x Turbofan, Booster: n.a., Orbiter: n.a.
Status	Inactive, n.a. - n.a., n.a.
Image / Data Source	http://www.astronautix.com
E-mail	eastronautica@hotmail.com

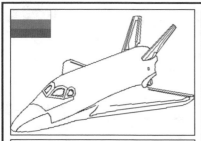

The LKS consists of a reusable winged spacecraft forwarded of an expendable payload section. The orbiter is launched on top of a Chelomei Proton booster.

LKS (Booster: Proton) ★★☆☆☆	
Developer	Chelomei
Type	Orbital
Number of Stages	4 (RKT-e: 3, ORB-r: 1)
Launch / Landing	Vertical / Horizontal
Payload	4,5 Mg
Launch Mass	719 Mg (Proton: 694 Mg, Orbiter: 25 Mg)
Propulsion System	Proton: 11 x Rocket, Orbiter: n.a.
Status	Inactive, 1975 - 1991, Development (Canceled)
Image / Data Source	http://www.astronautix.com
E-mail	eastronautica@hotmail.com

The M-46 project is a military version of the Raketoplan.

M-46 ★☆☆☆☆	
Developer	Myasishchev Design Bureau
Type	Orbital
Number of Stages	2 (RKT-e: 1, ORB-r: 1)
Launch / Landing	Vertical / Vertical
Payload	n.a. (130 km)
Launch Mass	n.a. (Booster: n.a., Orbiter: n.a.)
Propulsion System	Booster: 4 x Rocket, Orbiter: 1 x Rocket
Status	n.a., 1958 - n.a., n.a.
Image / Data Source	http://sb.balancer.ru/russia/myasishchev
E-mail	n.a.

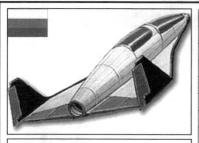

In 1958 the Soviet Air Force requested the development as quickly as possible of high-speed aerospace vehicles resulting in this system concept.

M-48 (Booster: R-7) ★★☆☆☆	
Developer	Myasishchev Design Bureau
Type	Orbital
Number of Stages	3 (RKT-e: 2, ORB-r: 1)
Launch / Landing	Horizontal / Horizontal
Payload	2 crew
Launch Mass	268 Mg (R-7: 263 Mg, Orbiter: 4,5 Mg)
Propulsion System	R-7: Rocket, Orbiter: 1 x Rocket
Status	Inactive, 1958 - 1960, n.a. (Canceled)
Image / Data Source	http://www.astronautix.com
E-mail	eastronautica@hotmail.com

Russia

MAKS (Booster: An-225)	★★★★☆
Developer	Molniya
Type	Orbital
Number of Stages	2 (PLN-r: 1, ORB-r: 1)
Launch / Landing	Horizontal / Horizontal
Payload	2 crew + 8,3 Mg (200 km / 51 deg)
Launch Mass	620 Mg (An-225: 345 Mg, Orbiter: 275 Mg)
Propulsion System	An-225: 6 x Turbofan, Orbiter: 2 x Rocket
Status	Inactive, 1988 - 1991, Development (Canceled)
Image / Data Source	http://www.astronautix.com
E-mail	eastronautica@hotmail.com

Development of MAKS was authorized but canceled in 1991. At that time, mock-ups of both the MAKS orbiter and the external tank had been finished.

MAKS-M (Booster: An-225)	★★★★☆
Developer	Molniya
Type	Orbital
Number of Stages	2 (PLN-r: 1, ORB-r: 1)
Launch / Landing	Horizontal / Horizontal
Payload	7,0 Mg (200 km / 0 deg)
Launch Mass	620 Mg (An-225: 345 Mg, Orbiter: 275 Mg)
Propulsion System	An-225: 6 x Turbofan, Orbiter: n.a.
Status	Inactive, 1988 - n.a., n.a.
Image / Data Source	http://www.astronautix.com
E-mail	eastronautica@hotmail.com

This system is a fully reusable unpiloted version of MAKS, similar to Interim HOTOL. MAKS-M requires new materials.

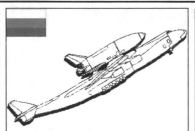

MiG 105-11 (Booster: Vostok)	★☆☆☆☆
Developer	MiG
Type	Orbital
Number of Stages	4 (RKT-e: 3, ORB-r: 1)
Launch / Landing	Vertical / Horizontal
Payload	1 crew
Launch Mass	285 Mg (Vostok: 281 Mg, MiG 105-11: 4,2 Mg)
Propulsion System	Vostok: 6 x Rocket, MiG 105-11: Turbojet + 3 x Rocket
Status	Inactive, n.a. - 1978, n.a. (Canceled)
Image / Data Source	http://www.astronautix.com
E-mail	eastronautica@hotmail.com

The MiG 105-11 is a flat-bottomed lifting body with a large upturned nose that earned it the nickname "Lapot" (wooden shoe).

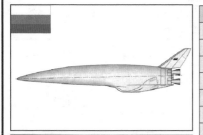

MiG 2000	★★★☆☆
Developer	MiG
Type	Orbital
Number of Stages	1
Launch / Landing	Horizontal / Horizontal
Payload	9,0 Mg (200 km / 51 deg)
Launch Mass	300 Mg
Propulsion System	Jet + Rocket
Status	Inactive, 1993 - n.a., n.a.
Image / Data Source	http://www.astronautix.com
E-mail	eastronautica@hotmail.com

MiG 2000 is a sled-launched orbital vehicle with air-breathing propulsion to Mach 5. The sled accelerates the launch vehicle to Mach 0,8.

Russia

MiG-31S (Booster: Mig-31S)	★★★★★
Developer	MiG
Type	Suborbital
Number of Stages	2 (PLN-r: 1, ORB-r: 1)
Launch / Landing	Horizontal / Horizontal
Payload	2 crew + 2 pax (130 km)
Launch Mass	n.a. (Mig-31S: 46 Mg, Orbiter: n.a.)
Propulsion System	Mig-31S: 2 x Turbofan, Orbiter: n.a.
Status	Inactive, n.a. - n.a., n.a.
Image / Data Source	http://www.astronautix.com
E-mail	eastronautica@hotmail.com

This is a commercial small satellite launcher variant with a Fakel OKB Micron missile capable of delivering a 100 kg payload into a 200 km orbit.

n.a.

MIGAKS	★★★☆☆
Developer	MiG
Type	Orbital
Number of Stages	2 (PLN-r: 1, ORB-r: 1)
Launch / Landing	Horizontal / Horizontal
Payload	12,6 Mg (200 km / 51 deg)
Launch Mass	420 Mg (Booster: n.a., Orbiter: n.a.)
Propulsion System	Booster: Turbojet + Ramjet, Orbiter: n.a.
Status	Inactive, 1993 - n.a., n.a.
Image / Data Source	http://www.astronautix.com
E-mail	eastronautica@hotmail.com

At Mach 6 the orbiter is separated from the first stage. The orbiter has 2000 km cross-range capability with landing on airfields with runways of 3500 m length.

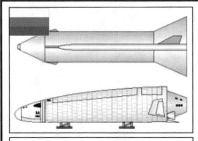

MTKVA (Booster: Vulkan)	★☆☆☆☆
Developer	Korolev
Type	Orbital
Number of Stages	3 (RKT-r: 2, ORB-r: 1)
Launch / Landing	Vertical / Horizontal
Payload	1 crew + 80,0 Mg
Launch Mass	3145 Mg (Vulkan: 2870 Mg, Orbiter: 275 Mg)
Propulsion System	Vulkan: 5 x Rocket, Orbiter: 1 x Rocket
Status	Inactive, 1974 - n.a., Concept
Image / Data Source	http://www.astronautix.com
E-mail	eastronautica@hotmail.com

The MTKVA is launched by the Vulkan launch vehicle into orbit. After completing its mission it can undertake a controlled re-entry from almost any orbit.

n.a.

MVKS	★★☆☆☆
Developer	Yakovlev
Type	Orbital
Number of Stages	1
Launch / Landing	n.a. / n.a.
Payload	n.a. (LEO)
Launch Mass	n.a.
Propulsion System	n.a.
Status	Inactive, 1986 - n.a., n.a.
Image / Data Source	http://www.astronautix.com
E-mail	eastronautica@hotmail.com

This system was required from government in reaction to X-30 project. MVKS provides effective and economic delivery to near Earth orbit.

Russia

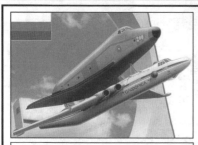

Myasishchev RLV is based on Atlant VM-T technologies. The VM-T Atlant is a plane adopted for transporting oversized cargo in an over wing container.

Myasishchev RLV (Booster: Atlant VM-T) ★★★☆☆

Developer	Myasishchev Design Bureau
Type	Orbital
Number of Stages	2 (PLN-r: 1, ORB-r: 1)
Launch / Landing	Horizontal / Horizontal
Payload	1,3 Mg
Launch Mass	182 Mg (Atlant VM-T: n.a., Orbiter: n.a.)
Propulsion System	Atlant VM-T: 4 x Turbojet, Orbiter: Rocket
Status	Active, n.a. - Present, n.a.
Image / Data Source	http://www.corbina.net/~kluka/Emz/Emz0-e.htm
E-mail	mdb@mail.sitek.ru

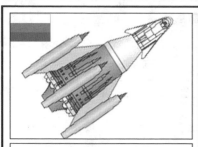

This system is a SSTO vehicle based on N1 Block A and described in RKK Energia's official history and in some detail in Peter James' 1974 book.

N1-MOK ★☆☆☆☆

Developer	Korolev
Type	Orbital
Number of Stages	1
Launch / Landing	Vertical / n.a.
Payload	90,0 Mg (450 km / 98 deg)
Launch Mass	1200 Mg
Propulsion System	20 x Rocket
Status	Inactive, 1974 - n.a., Concept
Image / Data Source	http://www.astronautix.com
E-mail	eastronautica@hotmail.com

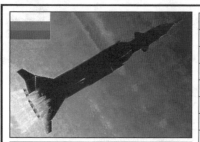

This is a semi-reusable vertically launched TSTO vehicle. This system uses a reusable flyback booster launched from a modular launch platform.

Norma ★★☆☆☆

Developer	n.a.
Type	Orbital
Number of Stages	3 (FBB-r: 1, RKT-e: 1, ORB-r: 1)
Launch / Landing	Vertical / Vertical
Payload	n.a.
Launch Mass	n.a. (Booster: n.a., Orbiter: n.a.)
Propulsion System	Booster: Rocket, Orbiter: n.a.
Status	n.a., 1997 - n.a., n.a.
Image / Data Source	http://www.astronautix.com
E-mail	eastronautica@hotmail.com

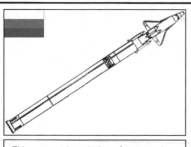

This concept is a design of a spaceplane smaller than Buran for space station crew rotation.

OK-M (Booster: Zenit) ★★☆☆☆

Developer	Molniya
Type	Orbital
Number of Stages	3 (RKT-e: 2, ORB-r: 1)
Launch / Landing	Vertical / Horizontal
Payload	6 crew + 3,5 Mg
Launch Mass	415 Mg (Zenit: 400 Mg, Orbiter: 15 Mg)
Propulsion System	Zenit: Rocket, Orbiter: Rocket
Status	n.a., 1980s - n.a., n.a.
Image / Data Source	http://www.astronautix.com
E-mail	eastronautica@hotmail.com

Russia

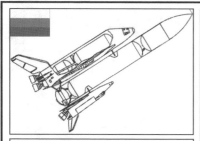

OK-M1 (Booster: RVK)	★★★☆☆
Developer	Molniya
Type	Orbital
Number of Stages	2 (FBB-r: 1, ORB-r: 1)
Launch / Landing	Vertical / Horizontal
Payload	8 crew + 7,2 Mg (250 km)
Launch Mass	832 Mg (RVK: 800 Mg, Orbiter: 32 Mg)
Propulsion System	RVK: 4 x Rocket, Orbiter: 2 x Rocket
Status	n.a., 1980s - n.a., n.a.
Image / Data Source	http://www.astronautix.com
E-mail	eastronautica@hotmail.com

The OK-M1 is one stage of a unique launch system, the MMKS which consists of the unmanned booster RVK and the expendable external tank PTO.

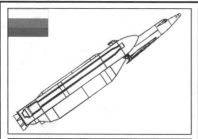

OK-M2 (Booster: Energia, Orbiter: OK-M2)	★★☆☆☆
Developer	Molniya
Type	Orbital
Number of Stages	2 (RKT-e: 1, ORB-r: 1)
Launch / Landing	Vertical / Horizontal
Payload	4 crew + 10,0 Mg (250 km)
Launch Mass	1060 Mg (Energia: 1030 Mg, OK-M2: 30 Mg)
Propulsion System	Energia: 8 x Rocket, OK-M2: 3 x Rocket
Status	Inactive, n.a. - n.a., n.a.
Image / Data Source	http://www.astronautix.com
E-mail	eastronautica@hotmail.com

The OK-M2 is positioned in a conventional manner on the nose of an Energia launch vehicle.

Orel V2	★★★☆☆
Developer	n.a.
Type	Orbital
Number of Stages	3 (FBB-r: 1, RKT-e: 1, ORB-r: 1)
Launch / Landing	Vertical / Horizontal
Payload	10,9 Mg (200 km / 51 deg)
Launch Mass	550 Mg (Booster: n.a., Orbiter: n.a.)
Propulsion System	Booster: Rocket, Orbiter: Rocket
Status	Inactive, 1997 - 1998, n.a. (Canceled)
Image / Data Source	http://www.astronautix.com
E-mail	eastronautica@hotmail.com

n.a.

This system has a flyback booster, an expendable second stage and a small manned spaceplane.

Orel V4	★★★☆☆
Developer	n.a.
Type	Orbital
Number of Stages	2 (FBB-r: 1, ORB-r: 1)
Launch / Landing	Vertical / Horizontal
Payload	10,0 Mg (200 km / 51 deg)
Launch Mass	n.a. (Booster: n.a., Orbiter: n.a.)
Propulsion System	Booster: n.a., Orbiter: n.a.
Status	Inactive, 1997 - n.a., n.a. (Canceled)
Image / Data Source	http://www.astronautix.com
E-mail	eastronautica@hotmail.com

n.a.

This is a fully reusable VTHL TSTO concept and was abandoned in favor of Orel V6.

Russia

n.a.

This vertical launched TSTO concept consists of a horizontal landing booster and a vertical landing orbiter. It has been abandoned in favor of Orel V6.

Orel V5	★★★☆☆
Developer	n.a.
Type	Orbital
Number of Stages	2 (FBB-r: 1, ORB-r: 1)
Launch / Landing	Vertical / Vertical
Payload	10,0 Mg (200 km / 51 deg)
Launch Mass	n.a. (Booster: n.a., Orbiter: n.a.)
Propulsion System	Booster: n.a., Orbiter: n.a.
Status	Inactive, 1997 - n.a., n.a. (Canceled)
Image / Data Source	http://www.astronautix.com
E-mail	eastronautica@hotmail.com

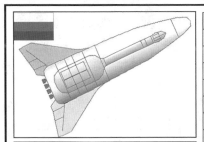

This system is a fully reusable VTHL SSTO launch vehicle and is the preferred long-term alternative of the Russian Orel launch vehicle study.

Orel V6	★★★☆☆
Developer	n.a.
Type	Orbital
Number of Stages	1
Launch / Landing	Vertical / Horizontal
Payload	10,0 Mg (200 km / 51 deg)
Launch Mass	930 Mg
Propulsion System	Rocket
Status	n.a., 1997 - n.a., n.a.
Image / Data Source	http://www.astronautix.com
E-mail	eastronautica@hotmail.com

n.a.

This concept was abandoned in favor of Orel V6 in 1998 due to engine reliability concerns. The engines consist of LOX and LH2 propellants.

Orel V7 RSSLV-2	★★★☆☆
Developer	n.a.
Type	Orbital
Number of Stages	1
Launch / Landing	Vertical / Vertical
Payload	10,0 Mg (200 km / 51 deg)
Launch Mass	1045 Mg
Propulsion System	Rocket
Status	Inactive, 1992 - 1998, n.a. (Canceled)
Image / Data Source	http://www.astronautix.com
E-mail	eastronautica@hotmail.com

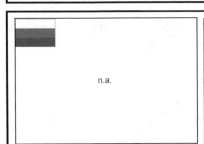

n.a.

This concept was abandoned in favor of Orel V6 in 1998 due to engine reliability concerns. It has tripropellant engines that are driven by LOX, Kerosene and LH2.

Orel V7 RSSLV-3	★★★☆☆
Developer	n.a.
Type	Orbital
Number of Stages	1
Launch / Landing	Vertical / Vertical
Payload	10,0 Mg (200 km / 51 deg)
Launch Mass	1058 Mg
Propulsion System	Rocket
Status	Inactive, 1992 - 1998, n.a. (Canceled)
Image / Data Source	http://www.astronautix.com
E-mail	eastronautica@hotmail.com

Russia

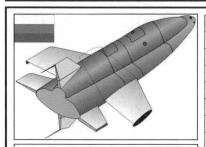

PKA (Booster: Vostok)	★☆☆☆☆
Developer	Tsybin
Type	Orbital
Number of Stages	4 (RKT-e: 3, ORB-r: 1)
Launch / Landing	Vertical / Horizontal
Payload	1 crew (300 km)
Launch Mass	275 Mg (Vostok: 270 Mg, Orbiter: 4,5 Mg)
Propulsion System	Vostok: 3 x Rocket, Orbiter: Rocket
Status	Inactive, 1959 - n.a., n.a. (Canceled)
Image / Data Source	http://www.astronautix.com
E-mail	eastronautica@hotmail.com

PKA is inserted into a 300 km altitude orbit by a Vostok launch vehicle. After 24 to 27 hours of flight the spacecraft starts a reentry.

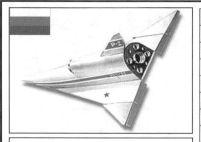

Raketoplan (Booster: Soyuz, Orbiter: R-2)	★☆☆☆☆
Developer	Chelomei
Type	Orbital
Number of Stages	4 (RKT-e: 3, ORB-r: 1)
Launch / Landing	Vertical / Horizontal
Payload	1 crew (290 km)
Launch Mass	308 Mg (Soyuz: 302 Mg, R-2: 6,3 Mg)
Propulsion System	Soyuz: Rocket, R-2: 5 x Rocket
Status	Inactive, 1959 - 1965, Development (Canceled)
Image / Data Source	http://www.astronautix.com
E-mail	eastronautica@hotmail.com

This system's solution for recovery was completely original and received a patent. The vehicle re-enters the atmosphere in a heat shield container.

Spiral OS (Booster: GSR, Booster: RB, Orbiter: OS)	★★★☆☆
Developer	MiG
Type	Orbital
Number of Stages	4 (PLN-r: 1, RKT-e: 2, ORB-r: 1)
Launch / Landing	Horizontal / Horizontal
Payload	1 crew (290 km)
Launch Mass	n.a. (GSR: n.a., RB: n.a., OS: n.a.)
Propulsion System	GSR: 4 × Turbo-ramjet, RB: Rocket, OS: n.a.
Status	Inactive, 1965 - n.a., Development
Image / Data Source	http://www.astronautix.com
E-mail	eastronautica@hotmail.com

The orbital spacecraft is a flat-bottomed lifting body, triangular in planform with a large upturned nose that earned it the nickname "Lapot".

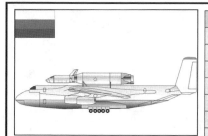

System 49 (Booster: An-124)	★★★☆☆
Developer	Molniya
Type	Orbital
Number of Stages	4 (PLN-r: 1, RKT-e: 2, ORB-r: 1)
Launch / Landing	Horizontal / Horizontal
Payload	1 crew + 4 Mg (391 km / 51 deg)
Launch Mass	630 Mg (An-124: 430 Mg, Booster: 187 Mg, Orbiter: 13 Mg)
Propulsion System	An-124: 4 x Turbofan, Booster: 4 x Rocket, Orbiter: n.a.
Status	Inactive, 1981 - n.a., n.a.
Image / Data Source	http://www.astronautix.com
E-mail	eastronautica@hotmail.com

This system can carry the payload to orbits from 120 to 1000 km altitude, and 45 to 94 degrees inclination.

Russia

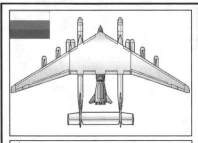

The System 49-M is a variation of the System 49 design concept, but with a larger carrier aircraft and a tripropellant single rocket stage.

System 49-M	★★★☆☆
Developer	Molniya
Type	Orbital
Number of Stages	3 (PLN-r: 1, RKT-e: 1, ORB-r: 1)
Launch / Landing	Horizontal / Horizontal
Payload	2 crew + 9,0 Mg (200 km / 51 deg)
Launch Mass	770 Mg (Boster: 400 Mg, Booster: 342 Mg, Orbiter: 28 Mg)
Propulsion System	Booster: n.a., Booster: 3 x Rocket, Orbiter: n.a.
Status	Inactive, 1981 - n.a., n.a. (Canceled)
Image / Data Source	http://www.astronautix.com
E-mail	eastronautica@hotmail.com

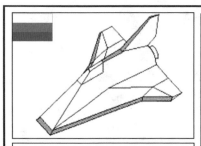

8 turbo-ramjets are supplemented by a rocket engine in order to achieve orbit for this vehicle concept.

Tu-2000	★☆☆☆☆
Developer	Tupolev
Type	Orbital
Number of Stages	1
Launch / Landing	Horizontal / Horizontal
Payload	10,0 Mg (200 km)
Launch Mass	260 Mg
Propulsion System	8 x Turbo-ramjet + 1 x Rocket
Status	n.a., 1986 - n.a., Development
Image / Data Source	http://www.astronautix.com
E-mail	eastronautica@hotmail.com

The crew ejects from the VKA-23 before landing and return separately to Earth by using a parachute.

VKA-23 Design 1	★☆☆☆☆
Developer	Myasishchev Design Bureau
Type	Orbital
Number of Stages	4 (RKT-e: 3, ORB-r: 1)
Launch / Landing	Vertical / Horizontal
Payload	1 crew + 0,7 Mg
Launch Mass	276 Mg (Vostok: 272 Mg, Orbiter: 3,5 Mg)
Propulsion System	Vostok: 6 x Rocket, Orbiter: n.a.
Status	Inactive, 1960 - n.a., n.a. (Canceled)
Image / Data Source	http://www.astronautix.com
E-mail	eastronautica@hotmail.com

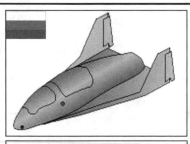

The crew ejects from the VKA-23 before landing and return separately to Earth by a parachute.

VKA-23 Design 2	★☆☆☆☆
Developer	Myasishchev
Type	Orbital
Number of Stages	4 (RKT-e: 3, ORB-r: 1)
Launch / Landing	Vertical / Horizontal
Payload	1 crew + 0,7 Mg
Launch Mass	276 Mg (Vostok: 272 Mg, Orbiter: 3,6 Mg)
Propulsion System	Vostok: 6 x Rocket, Orbiter: n.a.
Status	Inactive, 1960 - n.a., n.a. (Canceled)
Image / Data Source	http://www.astronautix.com
E-mail	eastronautica@hotmail.com

Russia

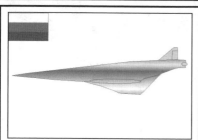

VKS	★☆☆☆☆
Developer	Korolev
Type	Orbital
Number of Stages	1
Launch / Landing	Horizontal / Horizontal
Payload	25,0 Mg (200 km / 51 deg)
Launch Mass	700 Mg
Propulsion System	Turboramjet + Rocket
Status	Inactive, 1986 - n.a., n.a. (Canceled)
Image / Data Source	http://www.astronautix.com
E-mail	eastronautica@hotmail.com

The VKS is designed as a hypersonic rocketplane with multi-regime engines. These engines are turboramjet with in-line rocket chambers.

n.a.

VKS-D (Booster: An-225)	★☆☆☆☆
Developer	Molniya
Type	Orbital
Number of Stages	2 (PLN-r: 1, ORB-r: 1)
Launch / Landing	Horizontal / Horizontal
Payload	6,0 Mg (200 km / 51 deg)
Launch Mass	620 Mg (An-225: 345 Mg, Orbiter: 275 Mg)
Propulsion System	An-225: 6 x Turbofan, Orbiter: n.a.
Status	Inactive, 1988 - n.a., n.a.
Image / Data Source	http://www.astronautix.com
E-mail	eastronautica@hotmail.com

The orbiter will be air launched from the An-225.

n.a.

VKS-G (Booster: Kholod)	★☆☆☆☆
Developer	Molniya
Type	Orbital
Number of Stages	2 (PLN-r: 1, ORB-r: 1)
Launch / Landing	Horizontal / Horizontal
Payload	21,0 Mg (200 km / 51 deg)
Launch Mass	750 Mg (Kholod: 550 Mg, Orbiter: 200 Mg)
Propulsion System	Kholod: Turboramjet, Orbiter: n.a.
Status	Inactive, 1988 - n.a., n.a. (Canceled)
Image / Data Source	http://www.astronautix.com
E-mail	eastronautica@hotmail.com

The orbiter is air-launched from the Kholod mother ship that is a modified Spiral 50-50.

n.a.

VKS-MD	★☆☆☆☆
Developer	Molniya
Type	Orbital
Number of Stages	2 (PLN-r: 1, ORB-r: 1)
Launch / Landing	Horizontal / Horizontal
Payload	n.a. (200 km / 51 deg)
Launch Mass	n.a. (Booster: n.a., Orbiter: 450 Mg)
Propulsion System	Booster: n.a., Orbiter: n.a.
Status	Inactive, 1988 - n.a., Development
Image / Data Source	http://www.astronautix.com
E-mail	eastronautica@hotmail.com

The orbiter is air-launched from the Geralk and Molniya twin-fuselage triplane.

Russia

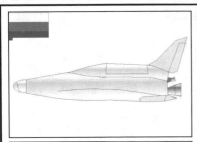

This vertically-launched SSTO design was considered too risky and the air-launched MAKS was selected instead.

VKS-O	★☆☆☆☆
Developer	Molniya
Type	Orbital
Number of Stages	1
Launch / Landing	Vertical / Vertical
Payload	8,5 Mg (200 km / 51 deg)
Launch Mass	550 Mg
Propulsion System	Rocket
Status	Inactive, 1988 - n.a., n.a.
Image / Data Source	http://www.astronautix.com
E-mail	eastronautica@hotmail.com

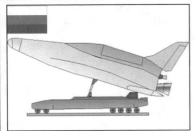

This system is a sled launch type resulting in a launch mass reduction but a velocity loss during lifting flight to orbit almost compensates this advantage.

VKS-R	★★☆☆☆
Developer	Molniya
Type	Orbital
Number of Stages	1
Launch / Landing	Horizontal / Horizontal
Payload	4,0 Mg (200 km / 51 deg)
Launch Mass	290 Mg
Propulsion System	Rocket
Status	Inactive, 1988 - n.a., n.a. (Canceled)
Image / Data Source	http://www.astronautix.com
E-mail	eastronautica@hotmail.com

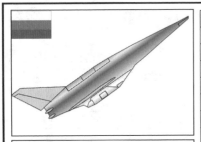

This vehicle is a horizontal take-off, delta winged SSTO launch vehicle with mixed rocket and ramjet propulsion.

VKS-RTO+ZhRD	★☆☆☆☆
Developer	Molniya
Type	Orbital
Number of Stages	1
Launch / Landing	Horizontal / Horizontal
Payload	2,0 Mg (200 km / 51 deg)
Launch Mass	770 Mg
Propulsion System	Ramjet + Rocket
Status	Inactive, 1998 - n.a., n.a.
Image / Data Source	http://www.astronautix.com
E-mail	eastronautica@hotmail.com

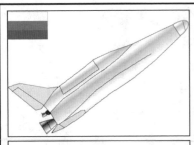

This system is similar to NASA's Shuttle-2 and RKK Energia's VKS. However this SSTO design was considered as too risky.

VKS-V	★☆☆☆☆
Developer	Molniya
Type	Orbital
Number of Stages	1
Launch / Landing	Vertical / Horizontal
Payload	7,5 Mg (200 km / 51 deg)
Launch Mass	550 Mg
Propulsion System	Rocket
Status	Inactive, 1988 - n.a., n.a. (Canceled)
Image / Data Source	http://www.astronautix.com
E-mail	eastronautica@hotmail.com

Russia

VKS-ZhRD+GPVRD	★☆☆☆☆
Developer	Molniya
Type	Orbital
Number of Stages	1
Launch / Landing	Horizontal / Horizontal
Payload	n.a. (200 km / 51 deg)
Launch Mass	770 Mg
Propulsion System	Rocket + Scramjet
Status	Inactive, 1988 - n.a., n.a.
Image / Data Source	http://www.astronautix.com
E-mail	eastronautica@hotmail.com

This system is a horizontal take-off delta winged SSTO launch vehicle with mixed rocket and scramjet propulsion.

United Kingdom

This vehicle is based on present technology. It will be enlarged to develop the Spacecab.

Ascender	★★☆☆☆
Developer	Bristol Spaceplanes
Type	Suborbital
Number of Stages	1
Launch / Landing	Horizontal / Horizontal
Payload	2 crew + 2 pax (100 km)
Launch Mass	4,5 Mg
Propulsion System	2 x Turbofan + 1 x Rocket
Status	Active, 1991 - Present, Development
Image / Data Source	http://www.bristolspaceplanes.com
E-mail	bsp@bristolspaceplanes.com

n.a.

Coopership can transport payloads to 100 km.

Coopership	★★☆☆☆
Developer	Coopership Industries
Type	Suborbital
Number of Stages	n.a.
Launch / Landing	n.a.
Payload	n.a. (100 km)
Launch Mass	n.a.
Propulsion System	n.a.
Status	Inactive, n.a. - n.a., n.a.
Image / Data Source	n.a.
E-mail	n.a.

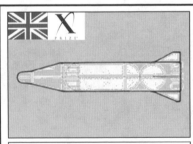

The main difference to other vehicles is that the whole vehicle will be recovered using standard parachutes.

Green Arrow	★☆☆☆☆
Developer	Flight Exploration
Type	Suborbital
Number of Stages	1
Launch / Landing	Vertical / Vertical (Parachute)
Payload	n.a. (100 km)
Launch Mass	20 Mg
Propulsion System	Rocket
Status	Active, n.a. - Present, n.a.
Image / Data Source	http://www.xprize.org
E-mail	peter.diamandis@xprize.org

The RB545 Air, LH2 and LOX rocket engine has been planned to be developed by Rolls-Royce. HOTOL takes off horizontally from a conventional runway.

HOTOL	★☆☆☆☆
Developer	British Aerospace
Type	Orbital
Number of Stages	1
Launch / Landing	Horizontal / Horizontal
Payload	n.a.
Launch Mass	250 Mg
Propulsion System	3 x Rocket
Status	Inactive, 1982 - 1990, n.a. (Canceled)
Image / Data Source	http://www.astronautix.com
E-mail	eastronautica@hotmail.com

United Kingdom

HOTOL 2 (Booster: An-225)	★★☆☆☆
Developer	British Aerospace
Type	Orbital
Number of Stages	2 (PLN-r: 1, ORB-r: 1)
Launch / Landing	Horizontal / Horizontal
Payload	n.a.
Launch Mass	595 Mg (An-225: 345 Mg, Orbiter: 250 Mg)
Propulsion System	An-225: 8 x Turbofan, Orbiter: 4 x Rocket
Status	Inactive, 1990 - n.a., Concept
Image / Data Source	http://www.astronautix.com
E-mail	eastronautica@hotmail.com

HOTOL 2 is a less ambitious scaled-back version of the original HOTOL. It is launched from the Ukrainian An-225 aircraft.

---

n.a.

Mustard	★☆☆☆☆
Developer	British Aircraft Corporation
Type	Orbital
Number of Stages	3 (FBB-r: 2, ORB-r: 1)
Launch / Landing	Vertical / Horizontal
Payload	n.a.
Launch Mass	424 Mg (Booster: 141 Mg, Booster: 141 Mg, Orbiter: 142 Mg)
Propulsion System	Booster: 1 x Rocket, Booster: 1 x Rocket, Orbiter: 1 x Rocket
Status	Inactive, 1964 - 1965, Concept
Image / Data Source	http://www.astronautix.com
E-mail	eastronautica@hotmail.com

This system is a winged reusable space shuttle using the Triamese concept. The orbiter is similar to the HL-10 vehicle.

---

n.a.

RAE Orbital Fighter	★☆☆☆☆
Developer	RAE
Type	Orbital
Number of Stages	2 (PLN-r: 1, ORB-r: 1)
Launch / Landing	Horizontal / Horizontal
Payload	n.a.
Launch Mass	n.a. (Booster: n.a., Orbiter: n.a.)
Propulsion System	Booster: Ramjet, Orbiter: n.a.
Status	Inactive, 1965 - n.a., n.a.
Image / Data Source	http://www.astronautix.com
E-mail	eastronautica@hotmail.com

This system is the Royal Air Craft Establishment Orbital Fighter proposal of the 1960s.

---

n.a.

RAE TSTO RLV	★☆☆☆☆
Developer	RAE
Type	Orbital
Number of Stages	2 (PLN-r: 1, ORB-r: 1)
Launch / Landing	Horizontal / Horizontal
Payload	n.a.
Launch Mass	n.a. (Booster: n.a., Orbiter: n.a.)
Propulsion System	Booster: Jet, Orbiter: Rocket
Status	Inactive, 1963 - n.a., Concept
Image / Data Source	http://www.astronautix.com
E-mail	eastronautica@hotmail.com

This system is the Royal Aircraft Establishment TSTO concept of the 1960s consisting of a hypersonic airbreathing first stage and a rocket powered second stage.

United Kingdom

Skylon	★☆☆☆☆
Developer	Reaction Engines Ltd.
Type	Orbital
Number of Stages	1
Launch / Landing	Horizontal / Horizontal
Payload	54 pax or 12,0 Mg
Launch Mass	n.a
Propulsion System	2 x Rocket
Status	Active, n.a. - Present, n.a.
Image / Data Source	http://www.gbnet.net/orgs/skylon
E-mail	richard.varvill@reactionengines.co.uk

This program is progressing well. This vehicle applies an unpiloted system.

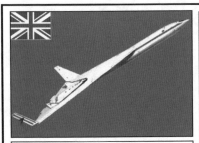

Spacebus	★☆☆☆☆
Developer	Bristol Spaceplanes
Type	Orbital
Number of Stages	2 (PLN-r: 1, ORB-r: 1)
Launch / Landing	Horizontal / Horizontal
Payload	50 pax (100 km)
Launch Mass	510 Mg (Booster: 400 Mg, Spacebus: 110 Mg)
Propulsion System	Booster: 4 x Turboramjet, Spacebus: Rocket
Status	Active, 1991 - Present, Concept
Image / Data Source	http://www.bristolspaceplanes.com
E-mail	bsp@bristolspaceplanes.com

Spacebus is an enlarged and mature development of Spacecab. These features include the use of jet plus rocket engines on the carrier aeroplane stage.

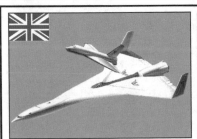

Spacecab	★☆☆☆☆
Developer	Bristol Spaceplanes
Type	Orbital
Number of Stages	2 (PLN-r: 1, ORB-r: 1)
Launch / Landing	Horizontal / Horizontal
Payload	2 crew + 6 pax
Launch Mass	n.a (Booster: n.a., Orbiter: n.a.)
Propulsion System	Booster: 4 x Turbojet, Spacecab: 2 x Rocket
Status	Active, 1990 - Present, Concept
Image / Data Source	http://www.bristolspaceplanes.com
E-mail	bsp@bristolspaceplanes.com

It is in effect an enlarged and refined Ascender.

Thunderbird	★★☆☆☆
Developer	Starchaser Industries
Type	Suborbital
Number of Stages	3 (RKT-e: 2, ORB-r: 1)
Launch / Landing	Vertical / Vertical (Parachute)
Payload	3 pax (100 km)
Launch Mass	20 Mg
Propulsion System	Rocket
Status	Active, n.a. - Present, Development
Image / Data Source	http://www.starchaser.co.uk
E-mail	ithc@starchaser.co.uk

This system contains three stage rocket vehicles. Their first manned flight has been planned for 2004.

United Kingdom

n.a.

Waverider 1960	★☆☆☆☆
Developer	Hawker Siddeley
Type	Orbital
Number of Stages	3 (RKT-e: 2, ORB-r: 1)
Launch / Landing	Vertical / Horizontal
Payload	n.a.
Launch Mass	n.a. (Booster: n.a., Orbiter: n.a.)
Propulsion System	Booster: n.a., Orbiter: n.a.
Status	Inactive, 1960s - n.a., n.a.
Image / Data Source	http://www.astronautix.com
E-mail	eastronautica@hotmail.com

The waverider spaceplane is positioned at the top of a LH2 and LOX second stage. Use of a nuclear upper stage would permit a British manned lunar landing by 1970.

n.a.

Waverider 1971	★☆☆☆☆
Developer	Hawker Siddeley
Type	Orbital
Number of Stages	2 (PLN-r: 1, ORB-r: 1)
Launch / Landing	Horizontal / Horizontal
Payload	4,0 Mg
Launch Mass	n.a. (Booster: n.a., Orbiter: n.a.)
Propulsion System	Booster: Jet, Orbiter: Rocket
Status	Inactive, 1960s - n.a., n.a.
Image / Data Source	http://www.astronautix.com
E-mail	eastronautica@hotmail.com

The Hawker Siddeley Waverider study of 1971 laid out a space vehicle with a airbreathing hypersonic first stage and a rocket propelled lifting body second stage.

USA

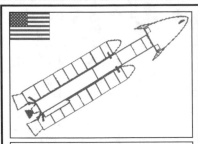

AACB means Aeronautics & Astronautics Coordinating Board. The Class 1 design uses a small 7 Mg reusable lifting-body spaceplane to carry the crew and payload.

AACB Class 1 (Booster: Titan 3M)	★★☆☆☆
Developer	USAF Study
Type	Orbital
Number of Stages	4 (RKT-e: 3, ORB-r: 1)
Launch / Landing	Vertical / Horizontal
Payload	2 crew + 4 pax (LEO)
Launch Mass	826 Mg (Titan 3M: 819 Mg, Class 1: 6,8 Mg)
Propulsion System	Titan 3M: Rocket, Class 1: n.a.
Status	Inactive, 1966 - n.a., Concept
Image / Data Source	http://www.astronautix.com
E-mail	eastronautica@hotmail.com

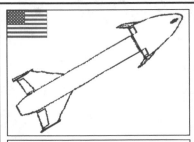

The AACB Class 2 concept is a fully reusable all-rocket TSTO vehicle.

AACB Class 2	★★☆☆☆
Developer	USAF Study
Type	Orbital
Number of Stages	2 (FBB-r: 1, ORB-r: 1)
Launch / Landing	Vertical / Horizontal
Payload	9,1 Mg (LEO)
Launch Mass	745 Mg (Booster: n.a., Orbiter: n.a.)
Propulsion System	Booster: Rocket, Orbiter: n.a.
Status	Inactive, 1966 - n.a., Concept
Image / Data Source	http://www.astronautix.com
E-mail	eastronautica@hotmail.com

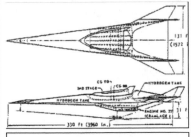

The AACB Class 3 concept is based on an airbreathing HTHL TSTO design investigated by Lockheed in 1965.

AACB Class 3	★☆☆☆☆
Developer	Lockheed
Type	Orbital
Number of Stages	2 (PLN-r: 1, ORB-r: 1)
Launch / Landing	Horizontal / Horizontal
Payload	15,9 Mg (LEO)
Launch Mass	438 Mg (Booster: 306 Mg, Orbiter: 132 Mg)
Propulsion System	Booster: Rocket, Orbiter: n.a.
Status	Inactive, 1965 - n.a., Concept
Image / Data Source	http://www.astronautix.com
E-mail	eastronautica@hotmail.com

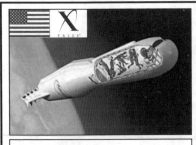

Advent is launched from water, so a launch site can be chosen. It is a complete reusable system.

Advent	★☆☆☆☆
Developer	American Advent
Type	Suborbital
Number of Stages	1
Launch / Landing	Vertical / Vertical (Parafoil)
Payload	6 pax (130 km)
Launch Mass	5,7 Mg
Propulsion System	1 x Rocket
Status	Inactive, 1996 – n.a., Development
Image / Data Source	http://www.ghg.net/jimakkerman
E-mail	jimakkerman@ghg.net

USA

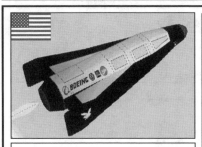

The Alpha CX-1A has three potential configuration options for the lower stage, VTHL, HTHL and VTVL.

Alpha CX-1A	★☆☆☆☆
Developer	World Aerospace
Type	Orbital
Number of Stages	2 (AAS-r: 1, ORB-r: 1)
Launch / Landing	n.a. / n.a.
Payload	1 crew + 2 pax
Launch Mass	n.a. (Booster: n.a., Orbiter: n.a.)
Propulsion System	Booster: n.a., Orbiter: n.a.
Status	Active, n.a. - n.a., Concept
Image / Data Source	http://www.spacefuture.com
E-mail	request@spacefuture.com

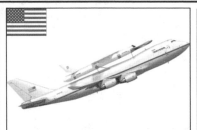

ALS is the Air Launch System under study by Boeing and Thiokol Propulsion.

ALS (Booster: Boeing 747)	★★★★☆
Developer	Boeing, Thiokol
Type	Orbital
Number of Stages	2 (PLN-r: 1, ORB-r: 1)
Launch / Landing	Horizontal / Horizontal
Payload	3,0 Mg (LEO)
Launch Mass	385 Mg (Boeing 747: 322 Mg, Orbiter: 63 Mg)
Propulsion System	Boeing 747: 4 x Turbofan, Orbiter: n.a.
Status	Active, n.a. - Present, n.a.
Image / Data Source	http://www.boeing.com
E-mail	wwwmail.boeing2@boeing.com

ALSV is the Boeing "Air-Launched Sortie Vehicle", which consists of a modified Boeing 747, a huge expendable tank and a rocket powered orbiter.

ALSV (Booster: Boeing 747-mod., Orbiter: ALSV)	★★★★☆
Developer	Boeing
Type	Orbital
Number of Stages	2 (PLN-r: 1, RKT-r: 1)
Launch / Landing	Horizontal / Horizontal
Payload	1,6 Mg (185 km)
Launch Mass	390 Mg (Boeing 747-mod.: 265 Mg, ALSV: 125 Mg)
Propulsion System	Boeing 747-mod.: 4 x Turbofan + 1 x Rocket, ALSV: 9 x Rocket
Status	Inactive, 1979 - 1982, n.a.
Image / Data Source	http://www.abo.fi/~mlindroo/SpaceLVs/Slides/sld053.htm
E-mail	n.a.

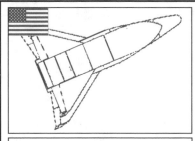

Advanced Manned Launch System (AMLS) is a fully reusable concept with an unmanned flyback booster and an external payload canister.

AMLS 1	★☆☆☆☆
Developer	NASA
Type	Orbital
Number of Stages	2 (FBB-r: 1, ORB-r: 1)
Launch / Landing	Vertical / Horizontal
Payload	9,1 Mg
Launch Mass	n.a. (Booster: n.a., Orbiter: n.a.)
Propulsion System	Booster: n.a., Orbiter: n.a.
Status	Inactive, 1989 - n.a., n.a.
Image / Data Source	http://www.abo.fi/~mlindroo/SpaceLVs/Slides/sld052.htm
E-mail	n.a.

USA

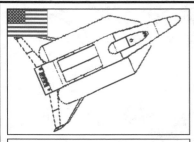

This Advanced Manned Launch System (AMLS) is a partially reusable concept with expendable hydrogen drop tanks.

AMLS 2	★☆☆☆☆
Developer	NASA
Type	Orbital
Number of Stages	2 (FBB-r: 1, ORB-r: 1)
Launch / Landing	Vertical / Horizontal
Payload	9,1 Mg
Launch Mass	n.a. (Booster: n.a., Orbiter: n.a.)
Propulsion System	Booster: n.a., Orbiter: n.a.
Status	Inactive, 1989 - n.a., n.a.
Image / Data Source	http://www.abo.fi/~mlindroo/SpaceLVs/Slides/sld052.htm
E-mail	n.a.

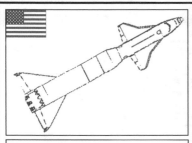

This Advanced Manned Launch System (AMLS) is a partially reusable concept with an expendable rocket stage.

AMLS 3	★★☆☆☆
Developer	NASA
Type	Orbital
Number of Stages	3 (FBB-r: 1, RKT-e: 1, ORB-r: 1)
Launch / Landing	Vertical / Horizontal
Payload	9,1 Mg
Launch Mass	n.a. (Booster: n.a., Orbiter: n.a.)
Propulsion System	Booster: n.a., Orbiter: n.a.
Status	Inactive, 1989 - n.a., n.a.
Image / Data Source	http://www.abo.fi/~mlindroo/SpaceLVs/Slides/sld052.htm
E-mail	n.a.

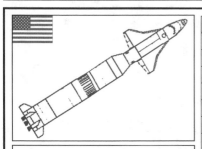

This Advanced Manned Launch System (AMLS) is a partially reusable concept with 2 expendable rocket stages.

AMLS 4	★★☆☆☆
Developer	NASA
Type	Orbital
Number of Stages	3 (RKT-e: 2, ORB-r: 1)
Launch / Landing	Vertical / Horizontal
Payload	9,1 Mg
Launch Mass	n.a. (Booster: n.a., Orbiter: n.a.)
Propulsion System	Booster: n.a., Orbiter: n.a.
Status	Inactive, 1989 - n.a., n.a.
Image / Data Source	http://www.abo.fi/~mlindroo/SpaceLVs/Slides/sld052.htm
E-mail	n.a.

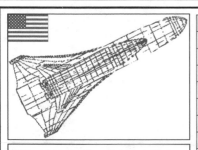

This Advanced Manned Launch System (AMLS) is a HTHL concept.

AMLS 5	★★★☆☆
Developer	NASA
Type	Orbital
Number of Stages	2 (PLN-1: 2, ORB-r: 1)
Launch / Landing	Horizontal / Horizontal
Payload	9,1 Mg
Launch Mass	n.a. (Booster: n.a., Orbiter: n.a.)
Propulsion System	Booster: n.a., Orbiter: n.a.
Status	Inactive, 1989 - n.a., n.a.
Image / Data Source	http://www.abo.fi/~mlindroo/SpaceLVs/Slides/sld052.htm
E-mail	n.a.

USA

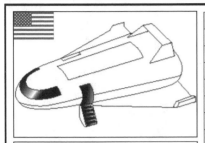

Angora	★☆☆☆☆
Developer	CFFC
Type	Orbital
Number of Stages	1
Launch / Landing	n.a. / Horizontal
Payload	40 pax or 15,0 Mg (LEO)
Launch Mass	464 Mg
Propulsion System	n.a.
Status	Active, n.a. - n.a., n.a.
Image / Data Source	http://www.spacefuture.com
E-mail	request@spacefuture.com

Angora is capable of carrying 40 people for two weeks in LEO aimed at the space tourism market. It is designed to use a metallic wire mesh parafoil for the re-entry.

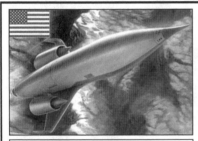

Argus	★☆☆☆☆
Developer	Georgia Institute of Technology
Type	Orbital
Number of Stages	1
Launch / Landing	Horizontal / Horizontal
Payload	9,1 Mg (LEO)
Launch Mass	256 Mg
Propulsion System	2 x Ramjet + Rocket
Status	Active, 1995 - n.a., Concept
Image / Data Source	http://www.ssdl.gatech.edu/main/index.html
E-mail	john.olds@ae.gatech.edu

This vehicle utilizes advanced vehicle technologies along with a magnetic levitation (Maglev) launch assist track.

AS&T Suborbital Aerospaceplane	★★☆☆☆
Developer	Andrews Space & Technology
Type	Suborbital
Number of Stages	2 (PLN-r: 1, ORB-r: 1)
Launch / Landing	Horizontal / Horizontal
Payload	6,4 Mg (120 km)
Launch Mass	n.a. (Booster: 123 Mg, Orbiter: n.a.)
Propulsion System	Booster: Turbojet, Orbiter: Rocket
Status	Active, n.a. - Present, n.a.
Image / Data Source	http://www.andrews-space.com/en/index.html
E-mail	info@andrews-space.com

This aerospaceplane will be operated similar to an aircraft, taking off horizontally with existing jet engines, and matching with the current air traffic control system.

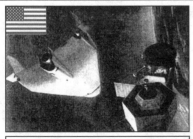

Astro	★☆☆☆☆
Developer	Douglas
Type	Orbital
Number of Stages	2 (FBB-r: 1, ORB-r: 1)
Launch / Landing	Vertical / Horizontal
Payload	16,9 Mg (555 km / 29 deg)
Launch Mass	408 Mg (Booster: 302 Mg, Orbiter: 106 Mg)
Propulsion System	Booster: 3 x Rocket, Orbiter: 3 x Rocket
Status	Inactive, 1962 - n.a., n.a.
Image / Data Source	http://www.astronautix.com
E-mail	eastronautica@hotmail.com

This system has been designed for launching space station crews and cargo in the 1960s.

USA

Astrocommuter (Booster: Saturn 1B)	★☆☆☆☆
Developer	Lockheed
Type	Orbital
Number of Stages	3 (RKT-e: 2, ORB-r: 1)
Launch / Landing	Vertical / Horizontal
Payload	7 pax
Launch Mass	n.a. (Saturn 1B: 590 Mg, Orbiter: n.a.)
Propulsion System	Saturn 1B: 9 x Rocket, Orbiter: n.a.
Status	Inactive, n.a. - n.a., n.a.
Image / Data Source	http://www.astronautix.com
E-mail	eastronautica@hotmail.com

Astrocommuter is Lockheed's space shuttle concept for space transportation to a space station in 1963.

n.a.

ATV	★★☆☆☆
Developer	NASA
Type	Orbital
Number of Stages	1
Launch / Landing	Vertical / Vertical
Payload	2 crew + 1,0 Mg (LEO)
Launch Mass	22 Mg
Propulsion System	n.a.
Status	Inactive, 1972 - n.a., n.a.
Image / Data Source	http://www.astronautix.com
E-mail	eastronautica@hotmail.com

George Detko at NASA's Marshall Space Flight Center designed this SSTO VTVL vehicle in 1972.

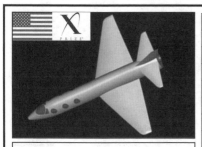

Aurora	★☆☆☆☆
Developer	Fundamental Technology Systems
Type	Suborbital
Number of Stages	1
Launch / Landing	Horizontal / Horizontal
Payload	0,3 Mg (105 km)
Launch Mass	5,7 Mg
Propulsion System	1 x Rocket
Status	Active, 2000 - Present, Development
Image / Data Source	http://www.funtechsystems.com
E-mail	n.a.

This is a horizontal all-rocket spaceplane design by Funtech.

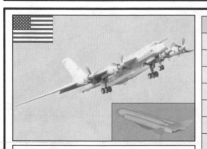

Bear Cub (Booster: Tu-95)	★★☆☆☆
Developer	Pan Aero
Type	Orbital
Number of Stages	3 (PLN-r: 2, ORB-r: 1)
Launch / Landing	Horizontal / Horizontal
Payload	1 crew + 3 pax (LEO)
Launch Mass	215 Mg (Tu-95: 185 Mg, Booster: 22 Mg, Orbiter: 8,4 Mg)
Propulsion System	Tu-95: 4 x Turboprop, Booster: 2 x Rocket, Orbiter: 1 x Rocket
Status	Active, n.a. - Present, n.a.
Image / Data Source	http://www.tour2space.com/index.htm
E-mail	panaero@tour2space.com

This concept uses a Tu-95 "Bear" bomber as first stage and is a near term, low-cost launch system.

USA

The Black Armadillo's propulsion system is a hydrogen peroxide monopropellant rocket engine fed from a single tank.

Black Armadillo	★☆☆☆☆
Developer	Armadillo Aerospace
Type	Suborbital
Number of Stages	1
Launch / Landing	Vertical / Vertical (Parachute)
Payload	0,3 Mg (107 km)
Launch Mass	6,4 Mg
Propulsion System	4 x Rockt
Status	Active, 2002 - Present, Development
Image / Data Source	http://www.armadilloaerospace.com
E-mail	pr@armadilloaerospace.com

Black Horse is a winged orbital launch vehicle using aerial refueling.

Black Horse	★☆☆☆☆
Developer	Mitchell Burnside Clapp
Type	Orbital
Number of Stages	1
Launch / Landing	Horizontal / Horizontal
Payload	2,5 Mg (200 km)
Launch Mass	84 Mg
Propulsion System	7 x Rocket
Status	Inactive, 1993 - n.a., n.a.
Image / Data Source	http://www.astronautix.com
E-mail	eastronautica@hotmail.com

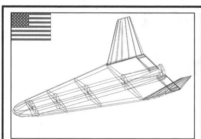

BLRV means Balloon Launched Return Vehicle. The vehicle's shape is a lifting-body type. It is launched by a balloon and lands horizontally.

BLRV (Orbiter: BLRV)	★☆☆☆☆
Developer	High Altitude Research Corporation
Type	Suborbital
Number of Stages	2 (BLN-e: 1, ORB-r: 1)
Launch / Landing	Vertical / Horizontal
Payload	n.a.
Launch Mass	n.a.
Propulsion System	Rocket
Status	Active, n.a. - n.a., Development
Image / Data Source	http://www.harcspace.com/2004
E-mail	harc@harcspace.com

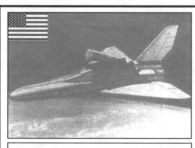

This is a sled-launched HTHL SSTO vehicle investigated by Boeing.

Boeing HTHL SSTO	★☆☆☆☆
Developer	Boeing
Type	Orbital
Number of Stages	1
Launch / Landing	Horizontal / Horizontal
Payload	113,4 Mg
Launch Mass	3438 Mg
Propulsion System	16 x Rocket
Status	Inactive, 1975 - n.a., n.a.
Image / Data Source	http://www.abo.fi/~mlindroo/SpaceLVs/Slides/sld041.htm
E-mail	n.a.

USA

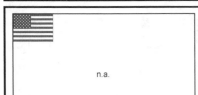

n.a.

This system can deliver components for a huge satellite solar power platform that NASA was promoting at that time.

Boeing SPS SSTO	★☆☆☆☆
Developer	Boeing
Type	Orbital
Number of Stages	1
Launch / Landing	Vertical / Vertical
Payload	227,0 Mg
Launch Mass	n.a.
Propulsion System	n.a.
Status	Inactive, 1977 - n.a., n.a.
Image / Data Source	http://www.astronautix.com
E-mail	eastronautica@hotmail.com

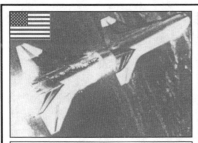

This is a fully reusable VTHL vehicle using a fly-back booster.

Boeing VTHL HLLV	★☆☆☆☆
Developer	Boeing
Type	Orbital
Number of Stages	2 (FBB-r: 1, ORB-r: 1)
Launch / Landing	Vertical / Horizontal
Payload	381,0 Mg
Launch Mass	9500 Mg
Propulsion System	n.a.
Status	Inactive, 1977 - n.a., n.a.
Image / Data Source	http://www.abo.fi/~mlindroo/SpaceLVs/Slides/sld045.htm
E-mail	n.a.

This is a fully reusable VTVL vehicle for solar power satellite heavy-lift missions.

Boeing VTVL HLLV	★☆☆☆☆
Developer	Boeing
Type	Orbital
Number of Stages	2 (RKT-r: 1, ORB-r: 1)
Launch / Landing	Vertical / Vertical
Payload	400,0 Mg (500 km)
Launch Mass	10 000 Mg
Propulsion System	14 x Rocket
Status	Inactive, 1976 - n.a., n.a.
Image / Data Source	http://www.abo.fi/~mlindroo/SpaceLVs/Slides/sld044.htm
E-mail	n.a.

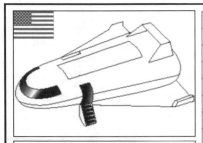

Calico is a short stay LEO vehicle capable of carrying 9 people aimed at the space tourism market. It is designed to use a metallic wire mesh parafoil to re-enter.

Calico	★☆☆☆☆
Developer	CFFC
Type	Orbital
Number of Stages	1
Launch / Landing	n.a. / Horizontal
Payload	9 pax or 1,8 Mg (LEO)
Launch Mass	52 Mg
Propulsion System	n.a.
Status	Active, n.a. - n.a., n.a.
Image / Data Source	http://www.spacefuture.com
E-mail	request@spacefuture.com

USA

Clipper Stormer MS-1	★☆☆☆☆
Developer	Space Clipper International
Type	Suborbital
Number of Stages	n.a.
Launch / Landing	n.a.
Payload	4 pax
Launch Mass	n.a.
Propulsion System	n.a.
Status	Inactive, n.a. - n.a., n.a.
Image / Data Source	http://www.edd.state.nm.us/PUBLICATIONS/strategic.pdf
E-mail	n.a.

n.a.

This suborbital vehicle has been proposed by Space Clipper International.

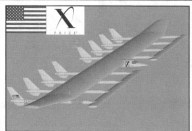

Condor-X	★★☆☆☆
Developer	Pan Aero
Type	Suborbital
Number of Stages	1
Launch / Landing	Horizontal / Horizontal
Payload	n.a. (100 km)
Launch Mass	24 Mg
Propulsion System	n.a.
Status	Active, n.a. - Present, n.a.
Image / Data Source	http://www.tour2space.com
E-mail	panaero@tour2space.com

This vehicle has a unique configuration. It uses the wing as parachute and moves the cabin under the wing in order to stabilize for descent.

Constellation	★★★☆☆
Developer	Marcus Aerospace
Type	Orbital
Number of Stages	1
Launch / Landing	Horizontal / Horizontal
Payload	0,1 Mg (500 km)
Launch Mass	5,3 Mg
Propulsion System	n.a.
Status	Active, n.a. - Present, n.a.
Image / Data Source	http://www.marcusaerospace.com/index.html
E-mail	info@marcusaerospace.com

n.a.

This vehicle is a fully reusable SSTO vehicle capable of launching a 75 kg microsatellite into a 500 km sun-synchronous orbit.

Convair Shuttlecraft	★★★☆☆
Developer	Convair
Type	Orbital
Number of Stages	2 (AAS: 1, ORB-r: 1)
Launch / Landing	Vertical / Horizontal
Payload	n.a.
Launch Mass	n.a. (Booster: n.a., Orbiter: n.a.)
Propulsion System	Booster: n.a., Orbiter: n.a.
Status	Inactive, 1960s - n.a., n.a.
Image / Data Source	http://www.astronautix.com
E-mail	eastronautica@hotmail.com

The Convair is a winged shuttle concept proposed in early 1960s.

USA

Copper Canyon	★★☆☆☆
Developer	DARPA
Type	Orbital
Number of Stages	1
Launch / Landing	Horizontal / Horizontal
Payload	n.a.
Launch Mass	n.a.
Propulsion System	Jet + Rocket
Status	Inactive, 1984 - n.a., n.a.
Image / Data Source	http://www.astronautix.com
E-mail	eastronautica@hotmail.com

n.a.

Copper Canyon is a Defense Advanced Research Project Agency program in 1984 that studied an airbreathing SSTO concept.

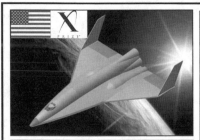

Cosmos Mariner	★☆☆☆☆
Developer	Lone Star Space Access
Type	Suborbital
Number of Stages	1
Launch / Landing	Horizontal / Horizontal
Payload	4 pax (100 km / 28 deg)
Launch Mass	62 Mg
Propulsion System	2 x Turbojet + 3 x Rocket
Status	Active, 1997 - Present, Concept
Image / Data Source	http://www.lonestarspace.com
E-mail	nlafave@lonestarspace.com

The Cosmos Mariner is a convenient, flexible, scalable and inexpensive space plane concept designed for operation at airport facilities.

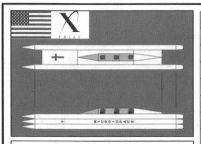

Crusader X	★★☆☆☆
Developer	Micro Space
Type	Suborbital
Number of Stages	1
Launch / Landing	Vertical / Vertical (Parafoil)
Payload	0,3 Mg (120 km)
Launch Mass	2,3 Mg
Propulsion System	Rocket
Status	Active, n.a. - Present, Development
Image / Data Source	http://www.micro-space.com
E-mail	n.a.

This vehicle uses a light core frame with seats and windshield resembling a bobsled or undersea sled.

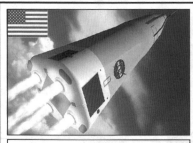

Delta Clipper DC-I	★★★☆☆
Developer	Douglas
Type	Orbital
Number of Stages	1
Launch / Landing	Vertical / Vertical
Payload	n.a.
Launch Mass	470 Mg
Propulsion System	1 x Rocket
Status	Inactive, 1985 - n.a., Development
Image / Data Source	http://www.friends-partners.ru/partners/mwade/lvs/dci.htm
E-mail	101326.3117@compuserve.com

This system is Douglas' VTVL SSTO vehicle.

USA

n.a.

This system has been proposed as an intermediate 1/2 scale test vehicle based on the DC-X and DC-Y orbital versions.

Delta Clipper DC-X2	★★★☆☆
Developer	Douglas
Type	Suborbital
Number of Stages	1
Launch / Landing	Vertical / Vertical
Payload	n.a.
Launch Mass	84 Mg
Propulsion System	Rocket
Status	Inactive, 1993 - n.a., n.a.
Image / Data Source	http://www.astronautix.com
E-mail	eastronautica@hotmail.com

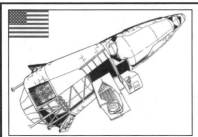

The ultimate goal of the Delta Clipper program is to make a prototype reusable VTVL SSTO vehicle. The DC-I Delta Clipper is a full version.

Delta Clipper DC-Y	★★★☆☆
Developer	Douglas
Type	Orbital
Number of Stages	1
Launch / Landing	Vertical / Vertical
Payload	4,5 Mg (300 km / 90 deg)
Launch Mass	470 Mg
Propulsion System	8 x Rocket
Status	Inactive, 1993 - n.a., n.a.
Image / Data Source	http://www.astronautix.com
E-mail	eastronautica@hotmail.com

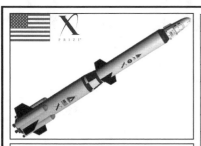

This vehicle is a three stage rocket system. Each stage will descent with a parachute and be reused.

Eagle	★★★☆☆
Developer	Vanguard Spacecraft
Type	Suborbital
Number of Stages	3 (RKT-r: 2, ORB-r: 1)
Launch / Landing	Vertical / Vertical (Parachute)
Payload	0,3 Mg (100 km)
Launch Mass	133 Mg
Propulsion System	Rocket
Status	Active, n.a. - Present, Development
Image / Data Source	http://www.xprize.org
E-mail	peter.diamandis@xprize.org

The lifting body will be towed to launch altitude behind another aircraft and after the separation rockets are ignited. It was canceled in favor for the KST Space Plane.

Eclipse Astroliner (Booster: Boeing 747)	★★★★☆
Developer	Kelly Space and Technology
Type	Suborbital
Number of Stages	2 (PLN-r: 1, ORB-r: 1)
Launch / Landing	Horizontal / Horizontal
Payload	40 pax (162 km)
Launch Mass	677 Mg (Boeing 747: 350 Mg, Eclipse Astroliner: 327 Mg)
Propulsion System	Boeing 747: 4 x Turbojet, Eclipse Astroliner: 3 x Rocket
Status	Inactive, 1998 - 2001, n.a. (Canceled)
Image / Data Source	http://www.kellyspace.com
E-mail	kstadmin@kellyspace.com

USA

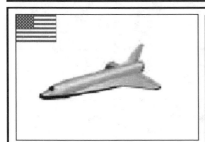

Eclipse Sprint	★★☆☆☆
Developer	Kelly Space and Technology
Type	Suborbital
Number of Stages	1
Launch / Landing	Horizontal / Horizontal
Payload	0,9 Mg (100 km)
Launch Mass	n.a.
Propulsion System	1 x Rocket
Status	n.a., 1993 - n.a., n.a.
Image / Data Source	http://www.kellyspace.com
E-mail	kstadmin@kellyspace.com

This vehicle is remote piloted and designed to meet the requirements of the academic and scientific communities for micro-gravity research and development.

n.a.

Exo-Clipper	★☆☆☆☆
Developer	Earth Space Transportation System
Type	Suborbital
Number of Stages	1
Launch / Landing	Horizontal / Horizontal
Payload	n.a.
Launch Mass	n.a.
Propulsion System	n.a.
Status	n.a., n.a. - n.a., n.a.
Image / Data Source	http://www.astronautix.com
E-mail	eastronautica@hotmail.com

This is a prototype air-breathing aerospace plane for suborbital flights and it is a X Prize competitor.

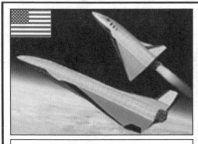

Explorer	★★☆☆☆
Developer	Marcus Aerospace
Type	Orbital
Number of Stages	2 (PLN-r: 1, ORB-r: 1)
Launch / Landing	Horizontal / Horizontal
Payload	0,5 Mg (400 km / 56 deg)
Launch Mass	377 Mg (Booster: 300 Mg, Orbiter: 77 Mg)
Propulsion System	Booster: n.a., Orbiter: Rocket
Status	Active, n.a. - Present, n.a.
Image / Data Source	http://www.marcusaerospace.com
E-mail	info@marcusaerospace.com

The two stages of the vehicle, the booster and the orbiter, will be developed sequentially rather than in parallel.

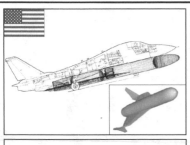

F-14 Space Transport (Booster: F-14)	★★★☆☆
Developer	Pan Aero
Type	Orbital
Number of Stages	2 (PLN-r: 1, ORB-r: 1)
Launch / Landing	Horizontal / Horizontal
Payload	0,1 Mg (800 km / 60 deg)
Launch Mass	40 Mg (F-14: 39,3 Mg, Orbiter: 0,7 Mg)
Propulsion System	F-14: 2 x Turbojet, Orbiter: 1 x Rocket
Status	Active, n.a. - Present, Concept
Image / Data Source	http://www.tour2space.com
E-mail	panaero@tour2space.com

The orbiter is mounted beneath a F-14. Pan Aero feels that one promising approach is to stage at Mach 2,5 and an altitude of 30 km.

USA

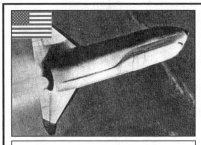

The Future Space Transportation System (FSTS) is a fully reusable VTHL TSTO concept.

FSTS	★☆☆☆☆
Developer	NASA
Type	Orbital
Number of Stages	2 (FBB-r: 1, ORB-r: 1)
Launch / Landing	Vertical / Horizontal
Payload	68 Mg (486 km / 31 deg)
Launch Mass	2223 Mg (Booster: 975 Mg, Orbiter:1248 Mg)
Propulsion System	Booster: 5 x Rocket, Orbiter: 7 x Rocket
Status	Inactive, 1981 - 1984, n.a.
Image / Data Source	http://www.abo.fi/~mlindroo/SpaceLVs/Slides/sld050.htm
E-mail	n.a.

n.a.

Edward Gomersall at NASA's Ames Research Center has produced a conservative design for an SSTO in 1970. The study is based on existing technology.

Gommersall	★☆☆☆☆
Developer	NASA
Type	Orbital
Number of Stages	1
Launch / Landing	Vertical / Vertical
Payload	100 Mg (LEO)
Launch Mass	n.a.
Propulsion System	Rocket
Status	Inactive, 1970 - n.a., Concept
Image / Data Source	http://www.astronautix.com
E-mail	eastronautica@hotmail.com

HSGS means Haynes Slipstream Gyro Saucer and is a X Prize competitor.

HSGS-711	★☆☆☆☆
Developer	Doug Haynes
Type	Suborbital
Number of Stages	1
Launch / Landing	Vertical / Vertical
Payload	n.a.
Launch Mass	n.a.
Propulsion System	2 x Ramjet
Status	Active, n.a. - Present, Development
Image / Data Source	www.blueridgeairlines.com
E-mail	dehas@sisna.com

This vehicle is a SSTO concept using LOX and LH2 ejector scramjet rocket-based combined cycle (RBCC) propulsion.

Hyperion	★☆☆☆☆
Developer	Georgia Institute of Technology
Type	Orbital
Number of Stages	1
Launch / Landing	Horizontal / Horizontal
Payload	9,1 Mg (LEO)
Launch Mass	363 Mg
Propulsion System	5 x Scramjet
Status	Inactive, 1996 - n.a., Concept
Image / Data Source	http://www.ssdl.gatech.edu/main/index.html
E-mail	john.olds@ae.gatech.edu

USA

Hyperion SSTO	★★☆☆☆
Developer	Douglas Missile & Space Systems
Type	Orbital
Number of Stages	1
Launch / Landing	Horizontal / Vertical
Payload	110 pax
Launch Mass	470 Mg
Propulsion System	Rocket
Status	Inactive, 1960s - n.a., n.a.
Image / Data Source	http://www.astronautix.com
E-mail	eastronautica@hotmail.com

This vehicle is planned to take off horizontally by using rail way acceleration and land vertically.

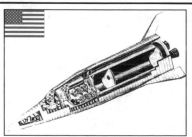

ILRV	★☆☆☆☆
Developer	McDonnell Douglas
Type	Orbital
Number of Stages	1
Launch / Landing	Vertical / Horizontal
Payload	22,7 Mg (550 km / 55 deg)
Launch Mass	1324 Mg
Propulsion System	5 x Rocket
Status	Inactive, 1968 - n.a., n.a.
Image / Data Source	http://www.astronautix.com
E-mail	eastronautica@hotmail.com

The McDonnell Douglas ILRV design featured fold-out wings for improved low-speed lift-to-drag ratio during final descent and landing.

Ithacus	★☆☆☆☆
Developer	Douglas
Type	Orbital
Number of Stages	1
Launch / Landing	Vertical / Vertical
Payload	1200 pax or 450,0 Mg (185 km / 28 deg)
Launch Mass	6363 Mg
Propulsion System	Rocket
Status	Inactive, 1966 - n.a., Concept
Image / Data Source	http://www.astronautix.com
E-mail	eastronautica@hotmail.com

This system is an adaptation of Rombus as a 1200 soldier intercontinental troop transport.

K-1	★★★★★
Developer	Kistler Aerospace
Type	Orbital
Number of Stages	2 (RKT-r: 1, ORB-r: 1)
Launch / Landing	Vertical / Vertical (Parachute)
Payload	4,6 Mg (200 km / 45 deg)
Launch Mass	382 Mg
Propulsion System	4 x Rocket
Status	Active, 1998 - Present, Development
Image / Data Source	http://www.kistleraerospace.com
E-mail	info@kistleraero.com

This vehicle's key factor is to use proven technologies and simple design to get high reliability.

USA

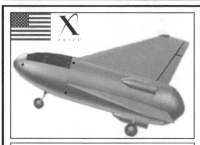

Kitten	★☆☆☆☆
Developer	CFFC
Type	Suborbital
Number of Stages	1
Launch / Landing	Horizontal / Horizontal (235 km)
Payload	2 pax + 1 crew
Launch Mass	3,2 Mg
Propulsion System	3 x Rocket
Status	Active, n.a. - Present, n.a.
Image / Data Source	http://www.xprize.org
E-mail	peter.diamandis@xprize.org

Kitten is Cerulean Freight Forwarding Company's entry for the X Prize.

KST Space Plane (Booster: Boeing 747)	★☆☆☆☆
Developer	Kelly Space and Technology, Vought Aircraft Industries
Type	Suborbital
Number of Stages	2 (PLN-r: 1, ORB-r: 1)
Launch / Landing	Horizontal / Horizontal
Payload	n.a.
Launch Mass	n.a. (Boeing 747: 350 Mg, Orbiter: n.a.)
Propulsion System	Boeing 747: 4 x Turbofan, Orbiter: n.a.
Status	Active, 2001 - Present, n.a.
Image / Data Source	http://www.kellyspace.com
E-mail	kstadmin@kellyspace.com

In 2001 Kelly Space & Technology and Vought Aircraft Industries have joined a project to design and produce this 2nd generation RLV system.

LB-X (Booster: F-4 Phantom)	★☆☆☆☆
Developer	Kelly Space and Technology
Type	Suborbital
Number of Stages	2 (PLN-r: 1, ORB-r: 1)
Launch / Landing	Horizontal / Horizontal
Payload	1 crew + 2 pax (100 km)
Launch Mass	n.a. (F-4 Phantom: 2,8 Mg, LB-X: n.a.)
Propulsion System	F-4 Phantom: 2 x Turbojet, LB-X: 1 x Rocket
Status	Active, 1993 - n.a., Development
Image / Data Source	http://www.kellyspace.com
E-mail	kstadmin@kellyspace.com

This system adopts KST's patented tow launch technology and utilizes a McDonnell Douglas F-4 Phantom. This vehicle was designed to meet X Prize requirements.

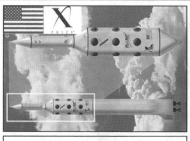

Liberator	★★☆☆☆
Developer	High Altitude Research Corporation
Type	Suborbital
Number of Stages	2 (RKT-r: 1, ORB-r: 1)
Launch / Landing	Vertical / Vertical (Parachute)
Payload	3 pax (113 km)
Launch Mass	4,5 Mg
Propulsion System	2 x Rocket
Status	Active, n.a. - Present, Development
Image / Data Source	http://www.harcspace.com/2004/index.php
E-mail	harc@harcspace.com

This vehicle is launched from ocean. Advantages of this method are longer launch windows, lower range costs and smoother FAA/AST licensing procedures.

USA

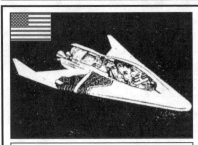

In June 1962, NASA's Marshall Space Flight Center awarded an 18-month contract worth $428 000 to Lockheed for studying a "Reusable Ten-Ton Orbital Carrier Vehicle".

Lockheed RTTOCV	★☆☆☆☆
Developer	Lockheed
Type	Orbital
Number of Stages	2 (PLN-r: 1, ORB-r: 1)
Launch / Landing	Horizontal / Horizontal
Payload	10 pax
Launch Mass	n.a. (Booster: n.a., Orbiter: n.a.)
Propulsion System	Booster: n.a., Orbiter: n.a.
Status	Inactive, 1963 - n.a., n.a.
Image / Data Source	http://www.astronautix.com
E-mail	eastronautica@hotmail.com

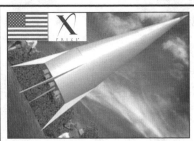

The propulsion section contains engines, liquid methane tanks, oxygen tanks and a helium pressurization system along with a main ascent control system.

Lucky Seven	★☆☆☆
Developer	Acceleration Engineering
Type	Suborbital
Number of Stages	1
Launch / Landing	Vertical / Vertical
Payload	1 crew + 2 pax (100 km)
Launch Mass	2,5 Mg
Propulsion System	Rocket
Status	Active, n.a. - Present, n.a.
Image / Data Source	http://www.xprize.org
E-mail	peter.diamandis@xprize.org

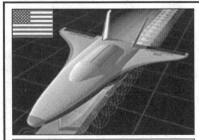

This concept uses an electro magnetic levitation rail as take-off assist system. The Marshall Space Flight Center tested this concept with a small vehicle model.

Magnetic Levitation System	★★★☆☆
Developer	NASA
Type	Orbital
Number of Stages	1
Launch / Landing	Horizontal / Horizontal
Payload	n.a.
Launch Mass	n.a.
Propulsion System	n.a.
Status	Active, n.a. - Present, n.a.
Image / Data Source	http://www1.msfc.nasa.gov
E-mail	dom.amatore@msfc.nasa.gov

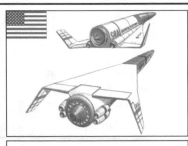

This system is an early TSTO shuttle study using storable propellants, Dynasoar's delta wing orbiter design and booster.

Martin Astrorocket	★★★☆☆
Developer	Martin
Type	Orbital
Number of Stages	2 (FBB-r: 1, ORB-r: 1)
Launch / Landing	Vertical / Horizontal
Payload	3 crew + 2,3 Mg (555 km)
Launch Mass	1134 Mg (Booster: 982 Mg, Orbiter: 152 Mg)
Propulsion System	Booster: 9 x Rocket + Turbojet, Orbiter: 1 x Rocket + Turbojet
Status	Inactive, 1962 - n.a., Concept
Image / Data Source	http://www.astronautix.com
E-mail	eastronautica@hotmail.com

USA

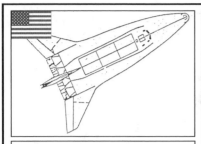

This concept is shown as Martin Marietta's parallel burn SSTO concept in the Hypersonic Aerospace Sizing Analysis (HASA) in 1988.

Martin Marietta SSTO	★☆☆☆☆
Developer	Martin Marietta
Type	Orbital
Number of Stages	1
Launch / Landing	Horizontal / Horizontal
Payload	29,5 Mg
Launch Mass	1055 Mg
Propulsion System	8 x Rocket
Status	Inactive, n.a. - n.a., n.a.
Image / Data Source	n.a.
E-mail	n.a.

This vehicle adopts a unique descent and landing system using a large dive brake to increase the drag coefficient of the vehicle.

MICHELLE-B	★☆☆☆☆
Developer	TGV Rockets
Type	Suborbital
Number of Stages	1
Launch / Landing	Vertical / Vertical
Payload	n.a. (104 km)
Launch Mass	28,0 Mg
Propulsion System	6 x Rocket
Status	Active, n.a. - Present, Development
Image / Data Source	http://www.tgv-rockets.com
E-mail	office@tgv-rockets.com

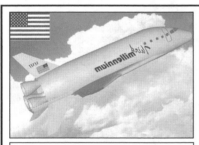

The orbiter stage follows an essential ballistic trajectory after separation at a high flight path angle. The orbiter wing is used only for re-entry.

Millennium Express a (Booster: An-22)	★☆☆☆☆
Developer	Pan Aero
Type	Orbital
Number of Stages	2 (PLN-r: 1, ORB-r: 1)
Launch / Landing	Horizontal / Horizontal
Payload	2 crew + 16 pax (450 km / 15 deg)
Launch Mass	n.a. (An-22: 250 Mg, Millennium Express: n.a.)
Propulsion System	An-22: 2 x Turboprop, Millennium Express: 3 x Rocket
Status	Active, n.a. - Present, n.a.
Image / Data Source	http://www.tour2space.com
E-mail	panaero@tour2space.com

n.a.

General Dynamics Space Systems Division's proposal for the 1990 SDIO competition is a VTOL SSTO concept named Millennium Express.

Millennium Express b	★☆☆☆☆
Developer	General Dynamics
Type	Orbital
Number of Stages	1
Launch / Landing	Vertical / Vertical
Payload	2 crew + 4,5 Mg (300 km / 90 deg)
Launch Mass	576 Mg
Propulsion System	1 x Rocket
Status	Inactive, 1990 - n.a., Concept
Image / Data Source	http://www.astronautix.com
E-mail	eastronautica@hotmail.com

USA

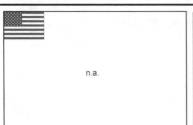

n.a.

By addition of many solid motors, up to 1588 Mg payload can be transported with a single launch.

MLLV	★☆☆☆☆
Developer	Boeing
Type	Orbital
Number of Stages	2 (RKT-e: 1, ORB-r: 1)
Launch / Landing	Vertical / Vertical
Payload	454,0 Mg (LEO)
Launch Mass	25 133 Mg
Propulsion System	24 x Rocket
Status	Inactive, 1969 - n.a., n.a.
Image / Data Source	http://www.astronautix.com
E-mail	eastronautica@hotmail.com

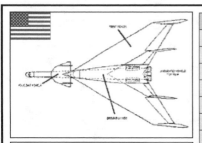

This system is North American Aviation's air-augmented VTVL SSTO RLV from 1963.

NAAA VTVL	★☆☆☆☆
Developer	North American
Type	Orbital
Number of Stages	1
Launch / Landing	Vertical / Vertical
Payload	454 Mg (LEO)
Launch Mass	13 607 Mg
Propulsion System	Ramjet + Rocket
Status	Inactive, 1963 - n.a., n.a.
Image / Data Source	http://www.astronautix.com
E-mail	eastronautica@hotmail.com

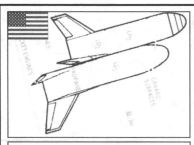

This system's name means North American Aviation's Reusable Ten Ton Orbital Carrier Vehicle. This system is similar to Lockheed's System 3 design.

NAA RTTOCV	★☆☆☆☆
Developer	North American
Type	Orbital
Number of Stages	2 (PLN-r: 1, ORB-r: 1)
Launch / Landing	Horizontal / Horizontal
Payload	12 pax (235 km / 28 deg)
Launch Mass	549 Mg (Booster: 538 Mg, Orbiter: 11 Mg)
Propulsion System	Booster: 3 x Rocket + Turbojet, Orbiter: 3 x Rocket
Status	Inactive, 1963 - n.a., n.a.
Image / Data Source	http://www.astronautix.com
E-mail	eastronautica@hotmail.com

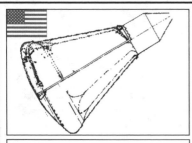

In early 1977 NASA decided to replace the VTVL recovery scheme to reduce the risk of vehicle loss and generated this system.

NASA SPS VTHL RLV	★☆☆☆☆
Developer	NASA
Type	Orbital
Number of Stages	2 (FBB-r: 1, ORB-r: 1)
Launch / Landing	Vertical / Horizontal
Payload	454,0 Mg (500 km)
Launch Mass	9526 Mg (Booster: 6500 Mg, Orbiter: 3026 Mg)
Propulsion System	Booster: 16 x Rocket, Orbiter: 14 x Rocket
Status	Inactive, 1977 - n.a., n.a.
Image / Data Source	http://www.abo.fi/~mlindroo/SpaceLVs/Slides/sld042.htm
E-mail	n.a.

USA

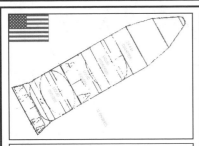

NASA SPS VTVL RLV ★☆☆☆☆

Developer	NASA
Type	Orbital
Number of Stages	2 (RKT-r: 1, ORB-r: 1)
Launch / Landing	Vertical / Vertical
Payload	700,0 Mg (500 km)
Launch Mass	14 203 Mg
Propulsion System	36 x Rocket
Status	Inactive, 1976 - n.a., n.a.
Image / Data Source	http://www.abo.fi/~mlindroo/SpaceLVs/Slides/sld042.htm
E-mail	n.a.

This system was the initial NASA concept studied from 1976 for launching 700 Mg solar power station subassemblies into LEO.

NASA VTHL RLV 1 ★★☆☆☆

Developer	NASA
Type	Orbital
Number of Stages	1
Launch / Landing	Vertical / Horizontal
Payload	45,0 Mg
Launch Mass	1158 Mg
Propulsion System	5 x Rocket
Status	Inactive, 1978 - n.a., n.a.
Image / Data Source	http://www.astronautix.com
E-mail	eastronautica@hotmail.com

This system is NASA's VTHL winged orbital launch vehicle.

n.a.

NASA VTHL RLV 2 ★★☆☆☆

Developer	NASA
Type	Orbital
Number of Stages	1
Launch / Landing	Vertical / Horizontal
Payload	9 Mg
Launch Mass	688 Mg
Propulsion System	3 x Rocket
Status	Inactive, 1978 - n.a., n.a.
Image / Data Source	http://www.astronautix.com
E-mail	eastronautica@hotmail.com

This system is NASA's VTHL winged orbital launch vehicle.

n.a.

NASA VTVL RLV ★★☆☆☆

Developer	NASA
Type	Orbital
Number of Stages	1
Launch / Landing	Vertical / Vertical
Payload	150 Mg
Launch Mass	4094 Mg
Propulsion System	17 x Rocket
Status	Inactive, 1978 - n.a., n.a.
Image / Data Source	http://www.astronautix.com
E-mail	eastronautica@hotmail.com

This system is NASA's VTVL winged orbital launch vehicle.

USA

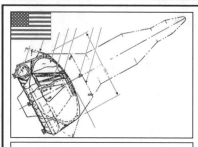

This system is a recoverable launch vehicle proposed by Krafft Ehricke at General Dynamics in early 1960s.

Nexus	★☆☆☆☆
Developer	General Dynamics
Type	Orbital
Number of Stages	1
Launch / Landing	Vertical / Vertical
Payload	450,0 Mg (LEO)
Launch Mass	21 820 Mg
Propulsion System	Rocket
Status	Inactive, 1962 - n.a., n.a.
Image / Data Source	http://www.astronautix.com
E-mail	eastronautica@hotmail.com

This vehicle is fueled with LOX during flight to reduce the take-off mass. This was an X Prize candidate. The project was stopped in favor of the Pioneer XP project.

Pathfinder	★★★☆☆
Developer	Pioneer Rocketplane
Type	Suborbital
Number of Stages	1
Launch / Landing	Horizontal / Horizontal
Payload	23 pax (130 km)
Launch Mass	n.a.
Propulsion System	2 x Turbojet + 1 x Rocket
Status	Inactive, n.a. - 2004, n.a.
Image / Data Source	http://www.rocketplane.com
E-mail	info@rocketplane.com

n.a.

The members of Aero Astro's team to develop this vehicle is only five.

PA-X2	★☆☆☆☆
Developer	Aero Astro
Type	Orbital
Number of Stages	1
Launch / Landing	Vertical / Vertical (Parachute)
Payload	n.a. (100 km)
Launch Mass	n.a.
Propulsion System	Rocket
Status	Active, n.a. - Present, Development
Image / Data Source	http://www.aeroastro.com
E-mail	info@aeroastro.com

This system design is a semi-SSTO ballistic VTVL launch vehicle with drop tanks that are separated on the way to orbit.

Pegasus VTVL	★☆☆☆☆
Developer	Douglas
Type	Orbital
Number of Stages	1
Launch / Landing	Vertical / Vertical
Payload	172 pax or 90,8 Mg (560 km / 28 deg)
Launch Mass	1520 Mg
Propulsion System	16 x Rocket
Status	Inactive, 1966 - n.a., n.a.
Image / Data Source	http://www.astronautix.com
E-mail	eastronautica@hotmail.com

USA

Phoenix C	★★☆☆☆
Developer	Pacific American Launch Systems
Type	Orbital
Number of Stages	1
Launch / Landing	Vertical / Vertical
Payload	n.a.
Launch Mass	180 Mg
Propulsion System	24 x Rocket
Status	Inactive, 1982 - 1988, n.a.
Image / Data Source	http://www.astronautix.com
E-mail	eastronautica@hotmail.com

This system is an unmanned cargo version. It has an aeroplug in place of an aerospike compared to earlier SSTO designs.

Phoenix E	★★☆☆☆
Developer	Pacific American Launch Systems
Type	Orbital
Number of Stages	1
Launch / Landing	Vertical / Vertical
Payload	20 pax
Launch Mass	205 Mg
Propulsion System	24 x Rocket
Status	Inactive, 1982 - 1988, n.a.
Image / Data Source	http://www.astronautix.com
E-mail	eastronautica@hotmail.com

This system is an excursion version for use as a Lunar or Mars lander.

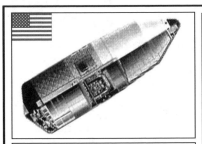

Phoenix L	★★☆☆☆
Developer	Pacific American Launch Systems
Type	Orbital
Number of Stages	1
Launch / Landing	Vertical / Vertical
Payload	n.a.
Launch Mass	31 Mg
Propulsion System	Rocket
Status	Inactive, 1982 - n.a., Concept
Image / Data Source	http://www.astronautix.com
E-mail	eastronautica@hotmail.com

Phoenix L is the only light cargo version VTVL SSTO vehicle among the Phoenix designs of the 1980s.

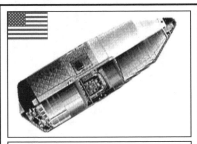

Phoenix LP	★★☆☆☆
Developer	Pacific American Launch Systems
Type	Orbital
Number of Stages	1
Launch / Landing	Vertical / Vertical
Payload	1 crew
Launch Mass	31 Mg
Propulsion System	Rocket
Status	Inactive, 1982 - n.a., Concept
Image / Data Source	http://www.astronautix.com
E-mail	eastronautica@hotmail.com

Phoenix LP is a light manned VTVL SSTO vehicle.

USA

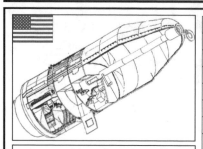

Phoenix M	★★☆☆☆
Developer	Pacific American Launch Systems
Type	Orbital
Number of Stages	1
Launch / Landing	Vertical / Vertical
Payload	n.a.
Launch Mass	n.a.
Propulsion System	Rocket
Status	Inactive, 1982 - 1988, Concept
Image / Data Source	http://www.astronautix.com
E-mail	eastronautica@hotmail.com

This is an intermediate version among the Phoenix concepts. Composite materials are used for cold structures and for propellant tanks.

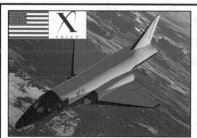

Pioneer XP	★★★★☆
Developer	Pioneer Rocketplane
Type	Suborbital
Number of Stages	1
Launch / Landing	Horizontal / Horizontal
Payload	2 crew + 2 pax (107 km)
Launch Mass	8 Mg
Propulsion System	2 x Turbojet + 2 x Rocket
Status	Active, n.a. - Present, n.a.
Image / Data Source	http://www.rocketplane.com
E-mail	info@rocketplane.com

This vehicle shares the operational flexibility advantages of the Pathfinder design.

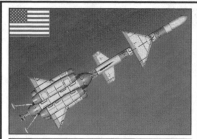

Pogo (Booster: Pogo, Orbiter: Pegasus)	★★☆☆☆
Developer	Alternative Accelerators
Type	Orbital
Number of Stages	2 (FBB-r: 1, ORB-r: 1)
Launch / Landing	Vertical / Horizontal
Payload	n.a.
Launch Mass	n.a. (Pogo: n.a., Pegasus: n.a.)
Propulsion System	Pogo: 5 x Ramjet, Pegasus: Rocket
Status	Active, n.a.- Present, Development
Image / Data Source	http://www.alt-accel.com
E-mail	glenn.olson@alt-accel.com

Pogo is the preliminary study of a reusable first stage for spacelift using existing aircraft jet engines.

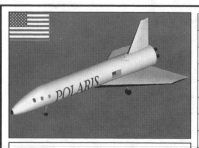

Polaris	★☆☆☆☆
Developer	Georgia Institute of Technology
Type	Suborbital
Number of Stages	1
Launch / Landing	Horizontal / Horizontal
Payload	3 pax
Launch Mass	n.a.
Propulsion System	3 x Rocket
Status	Inactive, n.a. - n.a., n.a.
Image / Data Source	http://www.ssdl.gatech.edu/main/index.html
E-mail	john.olds@ae.gatech.edu

This concept won a 1998 X Prize University design competition.

USA

Proteus is designed with long wings needed for efficient high altitude flights.

Proteus	★★★★☆
Developer	Scaled Composites
Type	Suborbital
Number of Stages	2 (PLN-r: 1, ORB: 1)
Launch / Landing	Horizontal / Horizontal
Payload	1 crew + 2 pax (130 km)
Weight	6,4 Mg (Booster: 3 Mg, Orbiter: 3,4 Mg)
Propulsion System	n.a.
Status	Inactive, n.a. - n.a., Development
Image / Data Source	http://www.scaled.com
E-mail	info@scaled.com

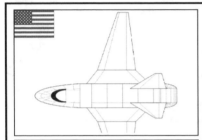

This vehicle is a potential competitor for the X Prize vehicles.

Raven	★☆☆☆☆
Developer	Beyond Earth Enterprises
Type	Suborbital
Number of Stages	1
Launch / Landing	Horizontal / Horizontal
Payload	1 crew + 2 pax (200 km)
Launch Mass	n.a.
Propulsion System	n.a.
Status	Active, n.a. - Present, Development
Image / Data Source	http://www.beyond-earth.com
E-mail	joe_latrell@beyond-earth.com

This system is Lockheed's Recoverable Booster System for orbital logistics. System I is uneconomical even if the crew capsule could be reused 10 times.

RBS System I (Booster: Saturn-IB, Orbiter: Apollo CSM)	★☆☆☆☆
Developer	Lockheed
Type	Orbital
Number of Stages	3 (RKT-e: 2, ORB-r: 1)
Launch / Landing	Vertical / Vertical
Payload	6 pax
Launch Mass	620 Mg (Saturn IB: 590 Mg, Apollo CSM: 30 Mg)
Propulsion System	Saturn IB: 9 x Rocket, Apollo CSM: 1 x Rocket
Status	Inactive, 1963 - n.a., n.a. (Canceled)
Image / Data Source	http://www.astronautix.com
E-mail	eastronautica@hotmail.com

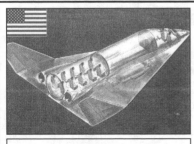

Lockheed proposed RBS System II instead of System I. System II's objective has been to replace the Apollo system.

RBS System II (Booster: Saturn-IB)	★☆☆☆☆
Developer	Lockheed
Type	Orbital
Number of Stages	3 (RKT-e: 2, ORB-r: 1)
Launch / Landing	Vertical / Horizontal
Payload	10 pax
Launch Mass	n.a. (Saturn IB: 590 Mg, Orbiter: n.a.)
Propulsion System	Saturn IB: 9 x Rocket, Orbiter: n.a.
Status	Inactive, 1963 - n.a., n.a. (Canceled)
Image / Data Source	http://www.astronautix.com
E-mail	eastronautica@hotmail.com

USA

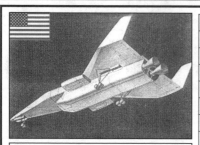

Lockheed proposed RBS System III as a fully reusable version. This system is similar to System IV but is a pure rocketplane version.

RBS System III	★☆☆☆☆
Developer	Lockheed
Type	Orbital
Number of Stages	3 (PLN-r: 2, ORB-r: 1)
Launch / Landing	Horizontal / Horizontal
Payload	10 pax + 3,0 Mg
Launch Mass	814 Mg (Booster: 647 Mg, Booster: 154 Mg, Orbiter: 13 Mg)
Propulsion System	Booster: Rocket, Booster: Rocket, Orbiter: n.a.
Status	Inactive, 1963 - n.a., n.a. (Canceled)
Image / Data Source	http://www.astronautix.com
E-mail	eastronautica@hotmail.com

This system is similar to System III but is an airbreathing HTHL ramjet rocketplane type.

RBS System IV	★☆☆☆☆
Developer	Lockheed
Type	Orbital
Number of Stages	3 (PLN-r: 2, ORB-r: 1)
Launch / Landing	Horizontal / Horizontal
Payload	10 pax + 3,0 Mg
Launch Mass	385 Mg (Booster: 218 Mg, Booster: 154 Mg, Orbiter: 13 Mg)
Propulsion System	Booster: Ramjet, Booster: Rocket, Orbiter: n.a.
Status	Inactive, 1963 - n.a., n.a. (Canceled)
Image / Data Source	http://www.astronautix.com
E-mail	eastronautica@hotmail.com

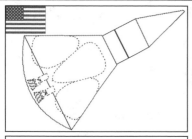

RITA C is a SSTO vehicle with nuclear, LOX and LH2 mixed cycle engines.

RITA C	★☆☆☆☆
Developer	Douglas
Type	Orbital
Number of Stages	1
Launch / Landing	Vertical / Vertical
Payload	454,5 Mg (325 km)
Launch Mass	4399 Mg
Propulsion System	4 x Rocket
Status	Inactive, 1963 - n.a., n.a.
Image / Data Source	http://www.astronautix.com
E-mail	eastronautica@hotmail.com

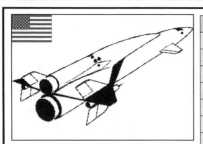

The Reusable Orbital Carrier (ROC) is a 1964 Lockheed study of a sled-launched HTHL TSTO vehicle.

ROC	★★★☆☆
Developer	Lockheed
Type	Orbital
Number of Stages	2 (PLN-r: 1, ORB-r: 1)
Launch / Landing	Horizontal / Horizontal
Payload	10 pax + 3,0 Mg
Launch Mass	453 Mg (Booster: n.a., Orbiter: n.a.)
Propulsion System	Booster: Rocket + Turbojet, Orbiter: Rocket
Status	Inactive, 1964 - n.a., n.a.
Image / Data Source	http://www.astronautix.com
E-mail	eastronautica@hotmail.com

USA

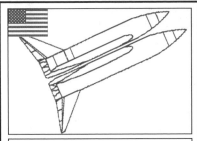

Rockwell VTHL RLV	★★★☆☆
Developer	North American Rockwell
Type	Orbital
Number of Stages	2 (FBB-r: 1, ORB-r: 1)
Launch / Landing	Vertical / Vertical
Payload	120,0 Mg (LEO)
Launch Mass	n.a. (Booster: n.a., Orbiter: n.a.)
Propulsion System	Booster: n.a., Orbiter: n.a.
Status	Inactive, 1980 - n.a., n.a.
Image / Data Source	http://www.abo.fi/~mlindroo/SpaceLVs/Slides/sld048.htm
E-mail	n.a.

This system is Rockwell's VTHL solar power satellite transportation concept in 1980.

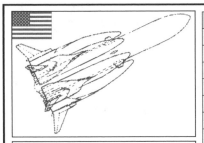

Rockwell VTVL RLV	★★☆☆☆
Developer	North American Rockwell
Type	Orbital
Number of Stages	2 (FBB-r: 1, ORB-r: 1)
Launch / Landing	Vertical / Vertical
Payload	n.a.
Launch Mass	6000 Mg
Propulsion System	Rocket
Status	Inactive, 1978 - n.a., n.a.
Image / Data Source	http://www.abo.fi/~mlindroo/SpaceLVs/Slides/sld048.htm
E-mail	n.a.

This is Rockwell's parallel-burn VTVL multistage concept consisting of 1 core stage and 6 flyback boosters.

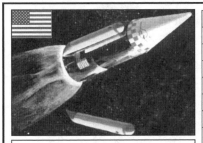

Rombus	★☆☆☆☆
Developer	Douglas
Type	Orbital
Number of Stages	1
Launch / Landing	Vertical / Vertical
Payload	450,0 Mg (185 km / 28 deg)
Launch Mass	6363 Mg
Propulsion System	36 x Rocket
Status	Inactive, 1964 - n.a., Concept
Image / Data Source	http://www.astronautix.com
E-mail	eastronautica@hotmail.com

This system design is a ballistic SSTO heavy lift launch vehicle with an expendable external tank.

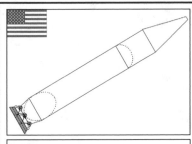

ROOST	★☆☆☆☆
Developer	Douglas
Type	Orbital
Number of Stages	1
Launch / Landing	Vertical / Vertical
Payload	4555,0 Mg (325 km)
Launch Mass	11 443 Mg
Propulsion System	4 x Rocket
Status	Inactive, 1963 - n.a., n.a.
Image / Data Source	http://www.astronautix.com
E-mail	eastronautica@hotmail.com

This system design is a reusable SSTO LH2 and LOX vehicle using conventional engines.

USA

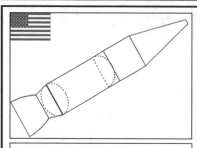

This system design is a reusable SSTO LH2 and LOX vehicle.

ROOST ISI	★☆☆☆☆
Developer	Douglas
Type	Orbital
Number of Stages	1
Launch / Landing	Vertical / Vertical
Payload	455,0 Mg (325 km)
Launch Mass	6763 Mg
Propulsion System	18 x Rocket
Status	Inactive, 1963 - n.a., n.a.
Image / Data Source	http://www.astronautix.com
E-mail	eastronautica@hotmail.com

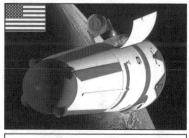

This vehicle takes off vertically like a conventional rocket. It is powered by a novel rotary engine burning liquid oxygen and jet fuel.

Roton C-9	★★☆☆☆
Developer	Rotary Rocket Company, Scaled Composites
Type	Orbital
Number of Stages	1
Launch / Landing	Vertical / Vertical
Payload	3,2 Mg (300 km / 50 deg)
Launch Mass	181 Mg
Propulsion System	Rocket
Status	Inactive, 1998 - 2001, Development
Image / Data Source	http://www.rotaryrocket.com
E-mail	n.a.

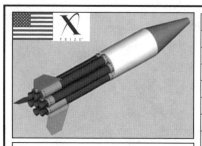

Power is provided by seven identical solid rocket engines - each is 58 cm in diameter and approximately 240 cm long. This vehicle is a X Prize competitor.

Rubicon	★☆☆☆☆
Developer	Space Transport Corporation
Type	Suborbital
Number of Stages	1
Launch / Landing	Vertical / Vertical (Parachute)
Payload	2 pax (100 km)
Launch Mass	2,3 Mg
Propulsion System	7 x Rocket
Status	Active, 2002 - Present, Development
Image / Data Source	http://www.space-transport.com
E-mail	pr@space-transport.com

This concept consists of an unpiloted spaceplane that uses a rocket-powered upper stage to deliver payloads to LEO or GTO.

SA-1 (Booster: SA-1)	★★☆☆☆
Developer	Space Access
Type	Suborbital
Number of Stages	2 (PLN-r: 1, ORB: 1)
Launch / Landing	Horizontal / Horizontal
Payload	15,0 Mg (LEO)
Launch Mass	n.a. (SA-1: n.a., Orbiter: n.a.)
Propulsion System	SA-1: Ramjet, Orbiter: Rocket
Status	Active, 1994 - Present, Development
Image / Data Source	http://www.spaceaccess.com
E-mail	betty@spaceaccess.com

USA

	SA-2	★★☆☆☆
Developer	Space Access	
Type	Suborbital	
Number of Stages	n.a.	
Launch / Landing	n.a. / n.a.	
Payload	n.a.	
Launch Mass	n.a.	
Propulsion System	Ramjet	
Status	Inactive, n.a. - n.a., n.a.	
Image / Data Source	http://www.orbireport.com/Linx/Startup.html	
E-mail	n.a.	

n.a.

This is a newer study based on the SA-1 results.

	Sabre Rocket	★★☆☆☆
Developer	Pan Aero	
Type	Suborbital	
Number of Stages	1	
Launch / Landing	Horizontal / Horizontal	
Payload	0,3 Mg (100 km)	
Launch Mass	12 Mg	
Propulsion System	2 x Turbojet + 7 x Rocket	
Status	Active, n.a. - Present, Development	
Image / Data Source	http://www.tour2space.com	
E-mail	panaero@tour2space.com	

This vehicle is a modification of the existing Sabre-40. In the phase of ballistic ascent it uses 7-clustered rocket engine.

	Salkeld RLV	★☆☆☆☆
Developer	Robert Salkeld	
Type	Orbital	
Number of Stages	1	
Launch / Landing	Vertical / Horizontal	
Payload	25,0 Mg (500 km / 55 deg)	
Launch Mass	2239 Mg	
Propulsion System	15 x Rocket	
Status	Inactive, 1965 - 1978, n.a.	
Image / Data Source	http://www.abo.fi/~mlindroo/SpaceLVs/Slides/sld039.htm	
E-mail	n.a.	

This is Robert Salkeld's fully reusable VTHL SSTO Shuttle II concept.

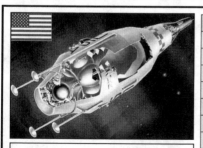

	SASSTO	★☆☆☆☆
Developer	Douglas	
Type	Orbital	
Number of Stages	1	
Launch / Landing	Vertical / Vertical	
Payload	2,8 Mg (185 km / 28 deg)	
Launch Mass	98 Mg	
Propulsion System	1 x Rocket	
Status	Inactive, 1967 - n.a., n.a.	
Image / Data Source	http://www.astronautix.com	
E-mail	eastronautica@hotmail.com	

This system is Bono's proposal for a first step toward a VTVL SSTO vehicle - a modified Saturn IVB with plug nozzle engine.

USA

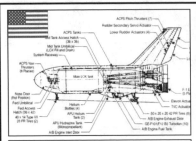

A winged recoverable Saturn IC stage has been considered instead of solid rocket boosters after the final shuttle design was selected.

Saturn Shuttle (Booster: Saturn IC)	★★☆☆☆
Developer	NASA
Type	Orbital
Number of Stages	2 (RKT-e: 1, ORB-r: 1)
Launch / Landing	Vertical / Horizontal
Payload	22,7 Mg
Launch Mass	3162 Mg (Saturn IC: 3037 Mg, Orbiter: 125 Mg)
Propulsion System	Saturn IC: 5 x Rocket, Orbiter: 5 x Rocket
Status	Inactive, 1974 - n.a., Concept
Image / Data Source	http://www.astronautix.com
E-mail	eastronautica@hotmail.com

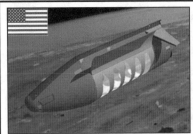

Large curved windows provide space tourists with breathtaking, unrestricted horizon-to-horizon views.

SC-1	★☆☆☆☆
Developer	Space Clipper International
Type	Suborbital
Number of Stages	1
Launch / Landing	Vertical / Vertical
Payload	2 crew + 10 pax (140 km)
Launch Mass	n.a.
Propulsion System	n.a.
Status	Inactive, n.a. - n.a., n.a.
Image / Data Source	http://www.spacelines.com/public/USLHomepage.htm
E-mail	n.a.

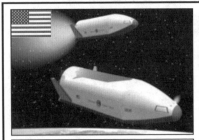

This second generation space plane is based on SC-1.

SC-2	★☆☆☆☆
Developer	Space Clipper International
Type	Orbital
Number of Stages	2 (AAS-r: 1, ORB-r: 1)
Launch / Landing	n.a. / n.a.
Payload	2 crew + 10 pax
Launch Mass	n.a. (Booster: n.a., Orbiter: n.a.)
Propulsion System	n.a.
Status	Active, n.a. - Present, n.a.
Image / Data Source	http://www.spacefuture.com
E-mail	request@spacefuture.com

NASA investigated a next generation launch system called "Shuttle II" which is envisioned as a fully reusable low cost piloted vehicle.

Shuttle II	★★★☆☆
Developer	NASA
Type	Orbital
Number of Stages	2 (FBB-r: 1, ORB-r: 1)
Launch / Landing	Vertical / Horizontal
Payload	9,1 Mg (400 km / 29 deg)
Launch Mass	923 Mg (Booster: 396 Mg, Orbiter: 527 Mg)
Propulsion System	Booster: 6 x Rocket, Orbiter: 5 x Rocket
Status	Inactive, 1985 - 1988, n.a.
Image / Data Source	http://www.abo.fi/~mlindroo/SpaceLVs/Slides/sld051.htm
E-mail	n.a.

USA

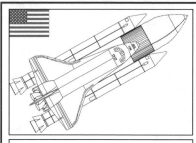

Shuttle ASRM	★★★☆☆
Developer	NASA
Type	Orbital
Number of Stages	2 (RKT-e: 1, ORB-r: 1)
Launch / Landing	Vertical / Horizontal
Payload	n.a.
Launch Mass	2100 Mg (Booster: 2001 Mg, Orbiter: 99 Mg)
Propulsion System	Booster: 2 x Rocket, Orbiter: 3 x Rocket
Status	Inactive, n.a. - 1993, Concept (Canceled)
Image / Data Source	http://www.astronautix.com
E-mail	eastronautica@hotmail.com

Shuttle ASRM uses Advanced Solid Rocket Motors (ASRM).

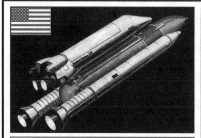

Shuttle C	★★★★☆
Developer	NASA
Type	Orbital
Number of Stages	2 (RKT-e: 1, ORB-r: 1)
Launch / Landing	Vertical / Horizontal
Payload	77,0 Mg (400 km / 28 deg)
Launch Mass	1967 Mg (Booster: 1931 Mg, Orbiter: 36 Mg)
Propulsion System	Booster: 2 × Rocket, Orbiter: 2 × Rocket
Status	Inactive, 1984 - n.a., n.a.
Image / Data Source	http://www.astronautix.com
E-mail	eastronautica@hotmail.com

Shuttle C is a Marshall Spaceflight Center design for replacement of the shuttle orbiter: A recoverable main engine pod.

Shuttle HCR	★★★☆☆
Developer	McDonnell Douglas, Martin Marietta
Type	Orbital
Number of Stages	2 (FBB-r: 1, ORB-r: 1)
Launch / Landing	Vertical / Horizontal
Payload	n.a.
Launch Mass	1977 Mg (Booster: 1634 Mg, Orbiter: 343 Mg)
Propulsion System	Booster: 14 x Rocket, Orbiter: 2 x Rocket
Status	Inactive,1969 - n.a., Concept
Image / Data Source	http://www.astronautix.com
E-mail	eastronautica@hotmail.com

n.a.

This system is McDonnell Douglas' and Martin Marietta's high cross-range shuttle proposal for phase B. It has swept wing booster and a delta wing orbiter.

Shuttle LCR	★★★☆☆
Developer	McDonnell Douglas, Martin Marietta
Type	Orbital
Number of Stages	2 (FBB-r: 1, ORB-r: 1)
Launch / Landing	Vertical / Horizontal
Payload	n.a.
Launch Mass	1834 Mg (Booster: 1512 Mg, Orbiter: 322 Mg)
Propulsion System	Booster: 13 x Rocket, Orbiter: 2 x Rocket
Status	Inactive,1969 - n.a., Concept
Image / Data Source	http://www.astronautix.com
E-mail	eastronautica@hotmail.com

n.a.

This system is McDonnell Douglas' and Martin Marietta's low cross-range shuttle proposal for phase B. It has swept wing booster and a straight wing orbiter.

USA

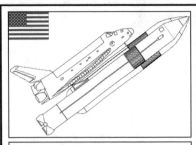

Shuttle LRB has Liquid Rocket Boosters in place of Solid Rocket Boosters.

Shuttle LRB	★★★☆☆
Developer	NASA
Type	Orbital
Number of Stages	2 (RKT-e: 1, ORB-r: 1)
Launch / Landing	Vertical / Horizontal
Payload	n.a.
Launch Mass	1575 Mg (Booster: 1451 Mg, Orbiter: 124 Mg)
Propulsion System	Booster: 8 x Rocket, Orbiter: 5 x Rocket
Status	Inactive, 1984 - n.a., n.a.
Image / Data Source	http://www.astronautix.com
E-mail	eastronautica@hotmail.com

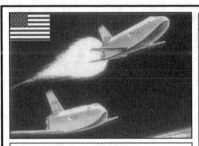

This system is Marshall Spaceflight Center's shuttle concept using a low cross-range stub-winged booster and orbiter.

Shuttle Phase A-1 DC-3	★★★★☆
Developer	NASA
Type	Orbital
Number of Stages	2 (FBB-r: 1, ORB-r: 1)
Launch / Landing	Vertical / Horizontal
Payload	n.a.
Launch Mass	999 Mg (Booster: 800 Mg, Orbiter: 199 Mg)
Propulsion System	Booster: 4 x Rocket, Orbiter: 1 x Rocket
Status	Inactive, 1970 - n.a., Concept
Image / Data Source	http://www.astronautix.com
E-mail	eastronautica@hotmail.com

This system is McDonnell Douglas' shuttle proposal for phase A. The first stage has delta wings and the second stage HL-10 is a lifting body type.

Shuttle Phase A-2 FR-3C (Orbiter: HL-10)	★☆☆☆☆
Developer	McDonnell Douglas
Type	Orbital
Number of Stages	2 (FBB-r: 1, ORB-r: 1)
Launch / Landing	Vertical / Horizontal
Payload	22,7 Mg
Launch Mass	1579 Mg (Booster: 1243 Mg, HL-10: 336 Mg)
Propulsion System	Booster: 10 x Rocket, HL-10: 2 x Rocket
Status	Inactive, 1969 - n.a., Concept
Image / Data Source	http://www.astronautix.com
E-mail	eastronautica@hotmail.com

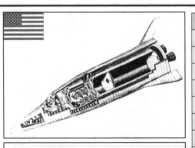

The specific launch cost is less because the drop-tank orbiter will carry a significantly larger payload than the drawbridge wing version shuttle.

Shuttle Phase A-3 Alternate	★☆☆☆☆
Developer	McDonnell Douglas
Type	Orbital
Number of Stages	2 (FBB-r: 1, ORB-r: 1)
Launch / Landing	Vertical / Horizontal
Payload	20 Mg (555 km / 55 deg)
Launch Mass	1601 Mg (Booster: 1331 Mg, Orbiter: 270 Mg)
Propulsion System	Booster: n.a., Orbiter: 2 x Rocket
Status	Inactive, 1969 - n.a., Concept
Image / Data Source	http://www.astronautix.com
E-mail	eastronautica@hotmail.com

USA

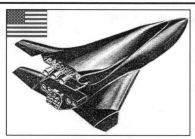

This system is Lockheed's shuttle proposal using the X-24B lifting body orbiter with delta-wing booster. First stage booster is a body-wing type.

Shuttle Phase A-4 LS112 ★☆☆☆☆

Developer	Lockheed
Type	Orbital
Number of Stages	2 (FBB-r: 1, ORB-r: 1)
Launch / Landing	Vertical / Horizontal
Payload	22,7 Mg
Launch Mass	1632 Mg (Booster: 1226 Mg, Orbiter: 406 Mg)
Propulsion System	Booster: 13 x Rocket, Orbiter: 3 x Rocket
Status	Inactive,1969 - n.a., Concept
Image / Data Source	http://www.astronautix.com
E-mail	eastronautica@hotmail.com

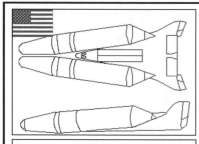

This system is Lockheed's alternate shuttle proposal and has a X-24B lifting body orbiter with a wrap-around external tank.

Shuttle Phase A-5 LS200 ★☆☆☆☆

Developer	Lockheed
Type	Orbital
Number of Stages	1
Launch / Landing	Vertical / Horizontal
Payload	n.a.
Launch Mass	1731 Mg
Propulsion System	9 x Rocket
Status	Inactive, 1971 - n.a., Concept
Image / Data Source	http://www.astronautix.com
E-mail	eastronautica@hotmail.com

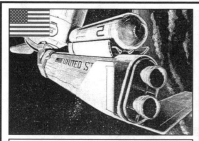

General Dynamics' shuttle proposal for phase A is an unwinged flat-bottom configuration booster and orbiter with V butterfly-tails.

Shuttle Phase A-6 FR-3A ★☆☆☆☆

Developer	Convair
Type	Orbital
Number of Stages	2 (FBB-r: 1, ORB-r: 1)
Launch / Landing	Vertical / Horizontal
Payload	22,7 Mg
Launch Mass	2558 Mg (Booster: 2170 Mg, Orbiter: 388 Mg)
Propulsion System	Booster: 15 x Rocket, Orbiter: 2 x Rocket
Status	Inactive, 1969 - n.a., Concept
Image / Data Source	http://www.astronautix.com
E-mail	eastronautica@hotmail.com

This is Martin Marietta's shuttle Phase A design. It has a X-24B type lifting body orbiter with an unique catamaran-configuration booster.

Shuttle Phase A-7 Spacemaster ★☆☆☆☆

Developer	Martin Marietta
Type	Orbital
Number of Stages	2 (FBB-r: 1, ORB-r: 1)
Launch / Landing	Vertical / Horizontal
Payload	22,7 Mg
Launch Mass	1587 Mg (Booster: 1224 Mg, Orbiter: 363 Mg)
Propulsion System	Booster: 14 x Rocket, Orbiter: 2 x Rocket
Status	Inactive, 1967 - n.a., Development
Image / Data Source	http://www.astronautix.com
E-mail	eastronautica@hotmail.com

USA

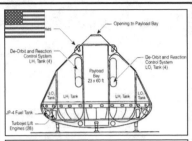

This system is Chrysler's ballistic SSTO alternative shuttle proposal. This has been the most detailed design study ever performed on a VTVL SSTO.

Shuttle Phase A-8 SERV	★☆☆☆☆
Developer	Chrysler
Type	Orbital
Number of Stages	1
Launch / Landing	Vertical / Vertical
Payload	52,8 Mg (300 km / 28 deg)
Launch Mass	2041 Mg
Propulsion System	1 x Rocket
Status	Inactive, 1971 - n.a., n.a. (Canceled)
Image / Data Source	http://www.astronautix.com
E-mail	eastronautica@hotmail.com

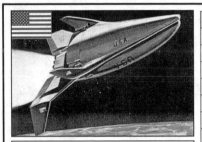

This is Boeing's and Lockheed's Shuttle proposal Phase B using delta wing. They considered another style with straight wings.

Shuttle Phase B-1	★☆☆☆☆
Developer	Boeing / Lockheed
Type	Orbital
Number of Stages	2 (FBB-r: 1, ORB-r: 1)
Launch / Landing	Vertical / Horizontal
Payload	n.a.
Launch Mass	n.a. (Booster: n.a., Orbiter: n.a.)
Propulsion System	Booster: 11 x Rocket, Orbiter: n.a.
Status	Inactive, 1970 - n.a., n.a. (Canceled)
Image / Data Source	http://www.abo.fi/~mlindroo/SpaceLVs/Slides/sld028.htm
E-mail	n.a.

This system is Rockwell's shuttle proposal phase A: A straight-wing low cross-range orbiter and booster.

Shuttle Phase B-2	★☆☆☆☆
Developer	North American Rockwell / General Dynamics
Type	Orbital
Number of Stages	2 (RKT-e: 1, ORB-r: 1)
Launch / Landing	Vertical / Horizontal
Payload	22,7 Mg
Launch Mass	2037 Mg (Booster: 1642 Mg, Orbiter: 395 Mg)
Propulsion System	Booster: 11 x Rocket, Orbiter: 2 x Rocket
Status	Inactive, 1969 - n.a., Concept
Image / Data Source	http://www.astronautix.com
E-mail	eastronautica@hotmail.com

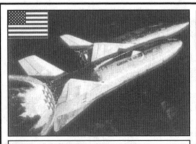

This system is Rockwell's and General Dynamics' shuttle proposal phase B: A delta wing high-cross range orbiter and booster.

Shuttle Phase B-3	★☆☆☆☆
Developer	North American Rockwell / General Dynamics
Type	Orbital
Number of Stages	2 (FBB-r: 1, ORB-r: 1)
Launch / Landing	Vertical / Horizontal
Payload	22,7 Mg
Launch Mass	2188 Mg (Booster: 1764 Mg, Orbiter: 424 Mg)
Propulsion System	Booster: 12 x Rocket, Orbiter: 2 x Rocket
Status	Inactive, 1970 - n.a., Concept
Image / Data Source	http://www.astronautix.com
E-mail	eastronautica@hotmail.com

USA

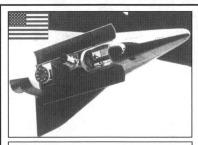

This system is McDonnell Douglas' and Martin Marietta's shuttle proposal phase B. The booster is derived from an alternative Spacemaster booster concept.

Shuttle Phase B-4	★☆☆☆☆
Developer	McDonnell Douglas / Martin Marietta
Type	Orbital
Number of Stages	2 (FBB-r: 1, ORB-r: 1)
Launch / Landing	Vertical / Horizontal
Payload	11,3 Mg (500 km / 55 deg)
Launch Mass	2107 Mg (Booster: 1717 Mg, Orbiter: 390 Mg)
Propulsion System	Booster: 12 x Rocket, Orbiter: 2 x Rocket
Status	Inactive, 1971 - n.a., n.a. (Canceled)
Image / Data Source	http://www.abo.fi/~mlindroo/SpaceLVs/Slides/sld031.htm
E-mail	n.a.

This system is Northrop Grumman's shuttle proposal Phase B.

Shuttle Phase B-5	★★☆☆☆
Developer	Northrop Grumman
Type	Orbital
Number of Stages	2 (FBB-r: 1, ORB-r: 1)
Launch / Landing	Vertical / Horizontal
Payload	n.a.
Launch Mass	n.a. (Booster: n.a., Orbiter: n.a.)
Propulsion System	Booster: n.a., Orbiter: n.a.
Status	Inactive, 1970 - n.a., n.a. (Canceled)
Image / Data Source	http://www.abo.fi/~mlindroo/SpaceLVs/Slides/sld032.htm
E-mail	n.a.

This system is Grumman's and Boeing's alternative proposal. The orbiter has drop tanks and the booster has delta wings.

Shuttle Phase B-6 H33	★☆☆☆☆
Developer	Northrop Grumman, Boeing
Type	Orbital
Number of Stages	2 (FBB-r: 1, ORB-r: 1)
Launch / Landing	Vertical / Horizontal
Payload	11,3 Mg (500 km / 55 deg)
Launch Mass	1964 Mg (Booster: 1490 Mg, Orbiter: 474 Mg)
Propulsion System	Booster: 12 x Rocket, Orbiter: 3 x Rocket
Status	Inactive, 1971 - n.a., Concept
Image / Data Source	http://www.astronautix.com
E-mail	eastronautica@hotmail.com

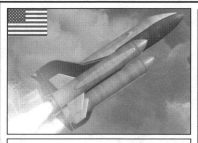

SLI 1 is one concept related to NASA's "Space Launch Initiative". Space Launch Initiative has been a program for next generation orbital plane developments.

SLI 1	★★★☆☆
Developer	Northrop Grumman
Type	Orbital
Number of Stages	2 (AAS: 1, ORB-r: 1)
Launch / Landing	n.a. / n.a.
Payload	n.a.
Launch Mass	n.a. (Booster: n.a., Orbiter: n.a.)
Propulsion System	Booster: n.a., Orbiter: n.a.
Status	Inactive, n.a. - n.a., n.a.
Image / Data Source	http://www.slinews.com
E-mail	kim.newton@msfc.nasa.gov

USA

SLI 2 is one concept related to NASA's "Space Launch Initiative" program.

SLI 2	★★★★☆
Developer	Northrop Grumman
Type	Orbital
Number of Stages	2 (PLN-r: 1, ORB-r: 1)
Launch / Landing	n.a. / n.a.
Payload	n.a.
Launch Mass	n.a. (Booster: n.a., Orbiter: n.a.)
Propulsion System	Booster: n.a., Orbiter: n.a.
Status	Inactive, n.a. - n.a., n.a.
Image / Data Source	http://xplanes.free.fr/x24/x24-24.htm, http://www.slinews.com
E-mail	kim.newton@msfc.nasa.gov

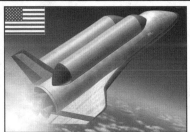

SLI 3 is one concept related to NASA's "Space Launch Initiative" program.

SLI 3	★★★☆☆
Developer	Northrop Grumman
Type	Orbital
Number of Stages	1
Launch / Landing	n.a. / n.a.
Payload	n.a.
Launch Mass	n.a.
Propulsion System	n.a.
Status	Inactive, n.a. - n.a., n.a.
Image / Data Source	http://www.slinews.com
E-mail	kim.newton@msfc.nasa.gov

SLI 4 is one concept related to NASA's "Space Launch Initiative" program.

SLI 4	★★★☆☆
Developer	Northrop Grumman
Type	Orbital
Number of Stages	3 (PLN-r: 2, ORB-r: 1)
Launch / Landing	n.a. / n.a.
Payload	n.a.
Launch Mass	n.a. (Booster: n.a., Booster: n.a., Orbiter: n.a.)
Propulsion System	Booster: n.a., Booster: n.a., Orbiter: n.a.
Status	Inactive, n.a. - n.a., n.a.
Image / Data Source	http://xplanes.free.fr/x24/x24-24.htm, http://www.slinews.com
E-mail	kim.newton@msfc.nasa.gov

SLI 5 is one concept related to NASA's "Space Launch Initiative" program.

SLI 5	★★★☆☆
Developer	Northrop Grumman
Type	Orbital
Number of Stages	1
Launch / Landing	n.a. / n.a.
Payload	n.a.
Launch Mass	n.a.
Propulsion System	n.a.
Status	Inactive, n.a. - n.a., n.a.
Image / Data Source	http://www.slinews.com
E-mail	kim.newton@msfc.nasa.gov

USA

SLI 6 is one concept related to NASA's "Space Launch Initiative" program.

SLI 6	★★★☆☆
Developer	Northrop Grumman
Type	Orbital
Number of Stages	3 (FBB-r: 2, ORB-r: 1)
Launch / Landing	Vertical / Horizontal
Payload	n.a.
Launch Mass	n.a. (Booster: n.a., Booster: n.a., Orbiter: n.a.)
Propulsion System	Booster: n.a., Booster: n.a., Orbiter: n.a.
Status	Inactive, n.a. - .n.a., n.a.
Image / Data Source	http://xplanes.free.fr/x24/x24-24.htm, http://www.slinews.com
E-mail	kim.newton@msfc.nasa.gov

SLI 7 is one concept related to NASA's "Space Launch Initiative" program.

SLI 7	★★★☆☆
Developer	Northrop Grumman
Type	Orbital
Number of Stages	3 (RKT-e: 2, ORB-r: 1)
Launch / Landing	Vertical / Horizontal
Payload	n.a.
Launch Mass	n.a. (Booster: n.a., Booster: n.a., Orbiter: n.a.)
Propulsion System	Booster: n.a., Booster: n.a., Orbiter: n.a.
Status	Inactive, n.a. - n.a., n.a.
Image / Data Source	http://www.slinews.com
E-mail	kim.newton@msfc.nasa.gov

SLI 8 is one concept related to NASA's "Space Launch Initiative" program.

SLI 8	★★★☆☆
Developer	Lockheed Martin
Type	Orbital
Number of Stages	3 (FBB-r: 2, ORB-r: 1)
Launch / Landing	n.a.
Payload	n.a.
Launch Mass	n.a. (Booster: n.a., Booster: n.a., Orbiter: n.a.)
Propulsion System	Booster: n.a., Booster: n.a., Orbiter: n.a.
Status	Inactive, n.a. - n.a., n.a.
Image / Data Source	http://xplanes.free.fr/x24/x24-24.htm, http://www.slinews.com
E-mail	kim.newton@msfc.nasa.gov

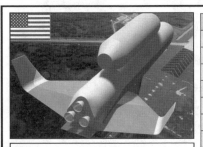

SLI 9 is one concept related to NASA's "Space Launch Initiative" program.

SLI 9	★★★☆☆
Developer	Lockheed Martin
Type	Orbital
Number of Stages	1
Launch / Landing	n.a. / Horizontal
Payload	n.a.
Launch Mass	n.a.
Propulsion System	n.a.
Status	Inactive, n.a. - n.a., n.a.
Image / Data Source	http://www.slinews.com
E-mail	kim.newton@msfc.nasa.gov

USA

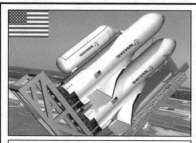

SLI 10 is one concept related to NASA's "Space Launch Initiative" program.

SLI 10	★★★☆☆
Developer	Boeing
Type	Orbital
Number of Stages	2 (FBB-r: 1, ORB-r: 1)
Launch / Landing	Vertical / Horizontal
Payload	n.a.
Launch Mass	n.a. (Booster: n.a., Orbiter: n.a.)
Propulsion System	Booster: n.a., Orbiter: n.a.
Status	Inactive, n.a. - n.a., n.a.
Image / Data Source	http://xplanes.free.fr/x24/x24-24.htm, http://www.slinews.com
E-mail	kim.newton@msfc.nasa.gov

SLI 11 is one concept related to NASA's "Space Launch Initiative" program.

SLI 11	★★★★★
Developer	Boeing
Type	Orbital
Number of Stages	3 (FBB-r: 2, ORB-r: 1)
Launch / Landing	Vertical / Horizontal
Payload	n.a.
Launch Mass	n.a. (Booster: n.a., Booster: n.a., Orbiter: n.a.)
Propulsion System	Booster: n.a., Booster: n.a., Orbiter: n.a.
Status	Inactive, n.a. - n.a., n.a.
Image / Data Source	http://www.slinews.com
E-mail	kim.newton@msfc.nasa.gov

SLI 12 is one concept related to NASA's "Space Launch Initiative" program.

SLI 12	★★★☆☆
Developer	NASA
Type	Orbital
Number of Stages	3 (FBB-r: 2, ORB-r: 1)
Launch / Landing	Vertical / Horizontal
Payload	n.a.
Launch Mass	n.a. (Booster: n.a., Booster: n.a., Orbiter: n.a.)
Propulsion System	Booster: n.a., Booster: n.a., Orbiter: n.a.
Status	Inactive, n.a. - n.a., n.a.
Image / Data Source	http://www.slinews.com
E-mail	kim.newton@msfc.nasa.gov

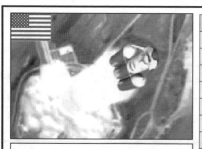

SLI 13 is one concept related to NASA's "Space Launch Initiative" program.

SLI 13	★★★☆☆
Developer	NASA
Type	Orbital
Number of Stages	3 (RKT-e: 2, ORB-r: 1)
Launch / Landing	Vertical / Horizontal
Payload	n.a.
Launch Mass	n.a. (Booster: n.a., Booster: n.a., Orbiter: n.a.)
Propulsion System	Booster: n.a., Booster: n.a., Orbiter: n.a.
Status	Inactive, n.a. - n.a., n.a.
Image / Data Source	http://www.slinews.com
E-mail	kim.newton@msfc.nasa.gov

USA

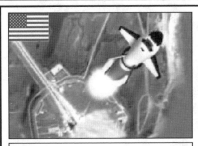

SLI 14 is one concept related to NASA's "Space Launch Initiative" program.

SLI 14	★★★☆☆
Developer	NASA
Type	Orbital
Number of Stages	3 (RKT-e: 2, ORB-r: 1)
Launch / Landing	Vertical / Horizontal
Payload	n.a.
Launch Mass	n.a. (Booster: n.a., Booster: n.a., Orbiter: n.a.)
Propulsion System	Booster: n.a., Booster: n.a., Orbiter: n.a.
Status	Inactive, n.a. - n.a., n.a.
Image / Data Source	http://www.slinews.com
E-mail	kim.newton@msfc.nasa.gov

SLI 15 is one concept related to NASA's "Space Launch Initiative" program.

SLI 15	★★★☆☆
Developer	NASA
Type	Orbital
Number of Stages	3 (RKT-e: 2, ORB-r: 1)
Launch / Landing	Vertical / Horizontal
Payload	n.a.
Launch Mass	n.a. (Booster: n.a., Booster: n.a., Orbiter: n.a.)
Propulsion System	Booster: n.a., Booster: n.a., Orbiter: n.a.
Status	Inactive, n.a. - n.a., n.a.
Image / Data Source	http://www.slinews.com
E-mail	kim.newton@msfc.nasa.gov

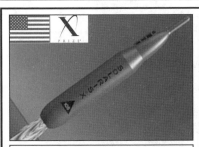

This vehicle has two separate recovery systems for safety and is launched from land or sea. Passengers use a re-entry capsule for emergency situations too.

Solaris-X	★★☆☆☆
Developer	Interorbital Systems
Type	Suborbital
Number of Stages	1
Launch / Landing	Vertical / Vertical (Parachute)
Payload	3 pax (161 km)
Launch Mass	6,8 Mg
Propulsion System	5 x Rocket
Status	Active, 1997 - Present, Development
Image / Data Source	http://www.interorbital.com/index.html
E-mail	ios@interorbital.com

This vehicle is a TSTO suborbital rocket concept. It is advertised in Incredible Adventures' space flight plan.

Space Cruiser (Booster: Sky Lifter)	★☆☆☆☆
Developer	Vela Technology Development Inc.
Type	Suborbital
Number of Stages	2 (PLN-r: 1, ORB-r: 1)
Launch / Landing	Horizontal / Horizontal
Payload	2 crew + 6 pax (250 km)
Launch Mass	30 Mg (Sky Lifter: 16 Mg, Space Cruiser: 14 Mg)
Propulsion System	Sky Lifter: 2 x Turbojet, Space Cruiser: 2 x Turbojet + 1 x Rocket
Status	Active, n.a. - Present, n.a.
Image / Data Source	http://www.spacetrans.com
E-mail	publicrelations@spacetrans.com

USA

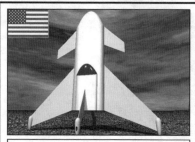

This vehicle is designed to be a manned reusable suborbital rocket.

Space Cub	★☆☆☆☆
Developer	David L. Burkhead
Type	Suborbital
Number of Stages	1
Launch / Landing	Vertical / Horizontal
Payload	4 pax (245 km)
Launch Mass	18 Mg
Propulsion System	3 x Rocket
Status	Active, n.a. - Present, Concept
Image / Data Source	http://www.sff.net/people/dburkhead/spacecub.htm
E-mail	dburkhead@sff.net

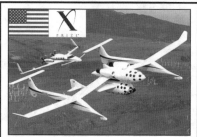

White Knight and SpaceShipOne have been realized and went through some flight tests. On June 21, 2004, SpaceShipOne reached space.

SpaceShipOne (Booster: White Knight)	★★★★★
Developer	Mojave Aerospace Ventures LLC., Scaled Composites
Type	Suborbital
Number of Stages	2 (PLN-r: 1, ORB-r: 1)
Launch / Landing	Horizontal / Horizontal
Payload	1 crew + 2 pax (115 km)
Launch Mass	7,7 Mg (White Knight: 4,1 Mg, SpaceShipOne: 3,6 Mg)
Propulsion System	White Knight: 2 x Turbojet, SpaceShipOne: 1 x Rocket
Status	Active, 1996 - Present, Realized (3 flights)
Image / Data Source	http://www.scaled.com
E-mail	info@scaled.com

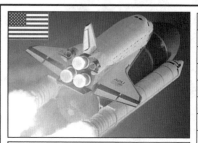

Space Shuttle is a manned reusable space system in operation which was designed to reduce the cost of space transport and replace all expendable launch vehicles.

Space Shuttle	★★★★★
Developer	NASA
Type	Orbital
Number of Stages	2 (RKT-e: 1, ORB-r: 1)
Launch / Landing	Vertical / Horizontal
Payload	7 crew + 24,4 Mg (204 km / 29 deg)
Launch Mass	2029 Mg (Booster: 1930 Mg, Orbiter: 99 Mg)
Propulsion System	Booster: 2 x Rocket, Orbiter: 3 x Rocket
Status	Active, 1981 - Present, Realized (115 flights)
Image / Data Source	http://www.nasa.gov
E-mail	comments@hq.nasa.gov

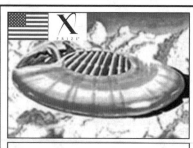

This rocket concept uses pulsejet technology.

Space Tourist	★☆☆☆☆
Developer	Discraft Corporation
Type	Suborbital
Number of Stages	1
Launch / Landing	Horizontal / Horizontal
Payload	6 pax (139 km)
Launch Mass	45 Mg
Propulsion System	Pulsejet
Status	Active, n.a. - Present, n.a.
Image / Data Source	http://www.xprize.org
E-mail	peter.diamandis@xprize.org

USA

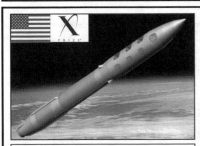

Spirit of Liberty	★★☆☆☆
Developer	American Astronautics
Type	Suborbital
Number of Stages	1
Launch / Landing	Vertical / Vertical (Parafoil)
Payload	7 pax (110 km)
Launch Mass	10 Mg
Propulsion System	1 x Rocket
Status	Active, n.a. - Present, n.a.
Image / Data Source	http://www.americanastronautics.com
E-mail	info@AmericanAstronautics.com

All flight functions are automatic. So passengers have no duties except to enjoy the ride.

n.a.

SSOAR	★☆☆☆☆
Developer	Earth/Space, Inc.
Type	Orbital
Number of Stages	1
Launch / Landing	Vertical / Vertical
Payload	n.a.
Launch Mass	n.a.
Propulsion System	1 x Rocket
Status	Inactive, 1976 - n.a., n.a. (Canceled)
Image / Data Source	http://www.astronautix.com
E-mail	eastronautica@hotmail.com

P. Seigler founded a company in 1976 to promote his design for a VTVL SSTO vehicle using a LOX and LH2 aerospike engine.

n.a.

SSX	★☆☆☆☆
Developer	Lockheed
Type	Orbital
Number of Stages	1
Launch / Landing	Vertical / Vertical
Payload	9,0 Mg (300 km / 28 deg)
Launch Mass	227 Mg
Propulsion System	Rocket
Status	Inactive, 1988 - n.a., n.a.
Image / Data Source	http://www.astronautix.com
E-mail	eastronautica@hotmail.com

This system is a VTVL SSTO design by Maxwell Hunter II at Lockheed in the late 1980s. SSX means SpaceShip eXperimental.

Star Bird I (Booster: Star Booster 200)	★★★★☆
Developer	Star Booster
Type	Orbital
Number of Stages	2 (FBB-r: 1, ORB-r: 1)
Launch / Landing	Vertical / Horizontal
Payload	8 crew (LEO)
Launch Mass	n.a. (Star Booster 200: n.a., Orbiter: n.a.)
Propulsion System	Star Booster 200: 1 x Rocket, Orbiter: 1 x Rocket
Status	Active, 1995 - Present, Development
Image / Data Source	http://www.starbooster.com
E-mail	starcraftboosters@yahoo.com

This system uses a unique crew escape module. The crew module will be separated from the vehicle and flies with its own propulsion system.

USA

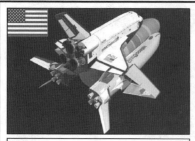

This concept is similar to the Space Shuttle System but using two Star Booster 750s instead of Solid Rocket Boosters.

Star Booster Space Shuttle (Booster: Star Booster 750) ★★★☆☆

Developer	Star Booster
Type	Orbital
Number of Stages	2 (FBB-r: 1, ORB-r: 1)
Launch / Landing	Vertical / Horizontal
Payload	n.a.
Launch Mass	n.a. (Star Booster 750: n.a., Orbiter: n.a.)
Propulsion System	Star Booster 750: 8 x Rocket, Orbiter: 3 x Rocket
Status	Inactive, n.a. - n.a., n.a.
Image / Data Source	http://www.starbooster.com
E-mail	starcraftboosters@yahoo.com

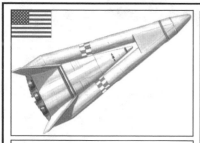

This concept is very similar to McDonnell Douglas' ILRV. This system is intended for space station crew transfers and resupply missions.

Starclipper ★☆☆☆☆

Developer	Lockheed
Type	Orbital
Number of Stages	1
Launch / Landing	Horizontal / Horizontal
Payload	11,3 Mg (555 km / 55 deg)
Launch Mass	326 Mg
Propulsion System	3 x Rocket
Status	Inactive, 1966 - n.a., n.a.
Image / Data Source	http://www.astronautix.com
E-mail	eastronautica@hotmail.com

The orbiter uses two Space Shuttle Main Engines (SSME) and two droppable tanks.

Star Eagle I (Booster: Star Booster 350) ★★★☆☆

Developer	Star Booster
Type	Orbital
Number of Stages	2 (FBB-r: 1, ORB-r: 1)
Launch / Landing	Vertical / Horizontal
Payload	4 crew + 44 pax (LEO)
Launch Mass	953 Mg (Star Booster 350: 379 Mg, Orbiter: 574)
Propulsion System	Star Booster 350: 4 x Rocket, Orbiter: 2 x Rocket
Status	Active, 1995 - Present, Development
Image / Data Source	http://www.starbooster.com
E-mail	starcraftboosters@yahoo.com

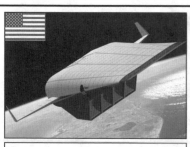

This vehicle has four LOX and LH2 ejector scramjet rocket-based combined-cycle engines.

Stargazer ★☆☆☆☆

Developer	Georgia Institute of Technology
Type	Orbital
Number of Stages	1
Launch / Landing	Horizontal / Horizontal
Payload	0,3 Mg (370 km)
Launch Mass	52,4 Mg
Propulsion System	Stargazer: 4 x Scramjet
Status	Inactive, 1999 - n.a., Concept
Image / Data Source	http://www.ssdl.gatech.edu
E-mail	john.olds@ae.gatech.edu

USA

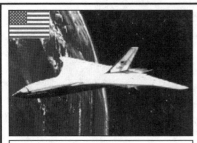

This concept is a HTHL SSTO concept with an estimated cost of $45 per kg payload to LEO.

Star Raker	★☆☆☆☆
Developer	Star-Raker Associates
Type	Orbital
Number of Stages	1
Launch / Landing	Horizontal / Horizontal
Payload	23,0 Mg (LEO)
Launch Mass	n.a.
Propulsion System	Ramjet + Rocket
Status	Inactive, n.a. - n.a., n.a.
Image / Data Source	http://www.astronautix.com
E-mail	eastronautica@hotmail.com

This concept consists of a reusable winged first stage vehicle and a low cost expendable upper stage.

Starsaber	★☆☆☆☆
Developer	Georgia Institute of Technology
Type	Suborbital
Number of Stages	2 (PLN-r: 1, ORB: 1)
Launch / Landing	Horizontal / Horizontal
Payload	0,1 Mg (250 km)
Launch Mass	n.a. (Starsaber: 76 Mg, Orbiter: n.a.)
Propulsion System	Starsaber: Ramjet, Orbiter: 1 x Rocket
Status	Active, 2002 - Present, Concept
Image / Data Source	http://www.ssdl.gatech.edu
E-mail	john.olds@ae.gatech.edu

This concept is USAF's program of 1980s that reached the test hardware stage but halted in favor of the X-30 National Aerospace plane.

TAV 1	★☆☆☆☆
Developer	Rockwell
Type	Orbital
Number of Stages	2 (PLN-r: 1, ORB-r: 1)
Launch / Landing	Horizontal / Horizontal
Payload	n.a.
Launch Mass	n.a. (Booster: n.a., Orbiter: n.a.)
Propulsion System	Booster: n.a., Orbiter: n.a.
Status	Inactive, 1980 - 1982, n.a. (Canceled)
Image / Data Source	http://www.astronautix.com
E-mail	eastronautica@hotmail.com

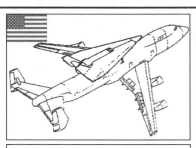

This is a fully reusable HTHL TSTO concept studied by General Dynamics.

TAV 2	★★☆☆☆
Developer	General Dynamics
Type	Orbital
Number of Stages	2 (PLN-r: 1, ORB-r: 1)
Launch / Landing	Horizontal / Horizontal
Payload	n.a.
Launch Mass	n.a. (Booster: n.a., Orbiter: n.a.)
Propulsion System	Booster: n.a., Orbiter: n.a.
Status	Inactive, 1982 - n.a., n.a.
Image / Data Source	http://www.abo.fi/~mlindroo/SpaceLVs/Slides/sld055.htm
E-mail	n.a.

USA

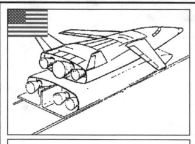

This is a fully reusable HTHL concept that takes off from a rocket-powered launch sled.

TAV 3	★☆☆☆☆
Developer	General Dynamics
Type	Orbital
Number of Stages	2 (SLD-r: 1, ORB-r:)
Launch / Landing	Horizontal / Horizontal
Payload	n.a.
Launch Mass	n.a
Propulsion System	n.a.
Status	Inactive, 1982 - n.a., n.a.
Image / Data Source	http://www.abo.fi/~mlindroo/SpaceLVs/Slides/sld055.htm
E-mail	n.a.

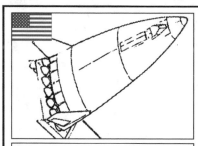

This is a fully reusable VTHL concept.

TAV 4	★☆☆☆☆
Developer	General Dynamics
Type	Orbital
Number of Stages	1
Launch / Landing	Vertical / Horizontal
Payload	n.a.
Launch Mass	n.a
Propulsion System	n.a.
Status	Inactive, 1982 - n.a., n.a.
Image / Data Source	http://www.abo.fi/~mlindroo/SpaceLVs/Slides/sld055.htm
E-mail	n.a.

This system is Lockheed's HTHL SSTO Trans Atmospheric Vehicle concept.

TAV 5	★★☆☆☆
Developer	Lockheed
Type	Orbital
Number of Stages	1
Launch / Landing	Horizontal / Horizontal
Payload	n.a.
Launch Mass	n.a.
Propulsion System	n.a.
Status	Inactive, 1984 - n.a., n.a.
Image / Data Source	http://www.abo.fi/~mlindroo/SpaceLVs/Slides/sld056.htm
E-mail	n.a.

This concept is capable of flights in and out of the atmosphere and speeds up to Mach 25.

TAV 6	★☆☆☆☆
Developer	McDonnell Douglas
Type	Orbital
Number of Stages	2 (RKT-r: 1, ORB-r: 1)
Launch / Landing	Vertical / Horizontal
Payload	4,5 Mg
Launch Mass	522 Mg
Propulsion System	n.a.
Status	Inactive, 1984 - n.a., n.a.
Image / Data Source	http://www.abo.fi/~mlindroo/SpaceLVs/Slides/sld057.htm
E-mail	n.a.

USA

	Tigre	★★☆☆☆
Developer	Kittyhawk Technologies	
Type	Suborbital	
Number of Stages	n.a.	
Launch / Landing	n.a. / Horizontal	
Payload	2 crew + 2 pax	
Launch Mass	7,5 Mg	
Propulsion System	n.a.	
Status	Inactive, n.a. - n.a., n.a.	
Image / Data Source	n.a.	
E-mail	n.a.	

This is a suborbital concept to transport 2 passengers.

	Triamese	★☆☆☆☆
Developer	General Dynamics	
Type	Orbital	
Number of Stages	3 (FBB-r: 2, ORB-r: 1)	
Launch / Landing	Vertical / Horizontal	
Payload	11,3 Mg (185 km / 28 deg)	
Launch Mass	518 Mg (Booster: n.a., Booster: n.a., Orbiter: n.a.)	
Propulsion System	Booster: n.a., Booster: n.a., Orbiter: 2 x Rocket	
Status	Inactive, 1968 - 1969, n.a. (Canceled)	
Image / Data Source	http://www.astronautix.com	
E-mail	eastronautica@hotmail.com	

General Dynamics proposed an ingenious Triamese concept for US Air Force's "Integral Launch & Re-entry Vehicle" program.

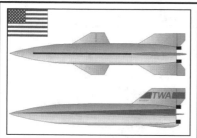

	Tsien Spaceplane 1949	★☆☆☆☆
Developer	Tsien Hsue-shen	
Type	Suborbital	
Number of Stages	1	
Launch / Landing	Vertical / Horizontal	
Payload	10 pax (160 km)	
Launch Mass	44 Mg	
Propulsion System	Rocket	
Status	Inactive, 1949 - n.a., Concept	
Image / Data Source	http://www.astronautix.com	
E-mail	eastronautica@hotmail.com	

This space plane was planned to carry ten passengers from New York to Los Angeles in 45 minutes.

	Venture Star	★★★★☆
Developer	Lockheed Martin	
Type	Orbital	
Number of Stages	1	
Launch / Landing	Vertical / Horizontal	
Payload	23,0 Mg (LEO)	
Launch Mass	1192 Mg	
Propulsion System	8 x Rocket	
Status	Inactive, 1996 - 1999, Development (Canceled)	
Image / Data Source	http://www.nasa.gov	
E-mail	carol.reukauf@dfrc.nasa.gov	

Venture Star is a RLV concept. NASA planned to prove the concept feasibility by X-33 experiments. The Venture Star project was stopped because of high costs in 1999.

USA

X-15 (Booster: B-52)	★★★★☆
Developer	NASA
Type	Suborbital
Number of Stages	2 (PLN-r: 1, ORB-r: 1)
Launch / Landing	Horizontal / Horizontal
Payload	1 crew (108 km)
Launch Mass	204 Mg (B-52: 174 Mg, X-15: 30 Mg)
Propulsion System	B-52: 8 x Turbofan, X-15: 2 x Rocket
Status	Inactive, 1954 - 1968, Realized (13 flights higher than 80 km)
Image / Data Source	http://www.x15.com/program.html
E-mail	comments@hq.nasa.gov

X-15 was designed to explore the challenge of flight at very high speeds and altitudes. This program was a joint effort of NASA, US Air Force and US Navy.

X-20 Dyna Soar (Booster: Titan 3C)	★☆☆☆☆
Developer	NASA
Type	Orbital
Number of Stages	4 (RKT-e: 3, ORB-r: 1)
Launch / Landing	Vertical / Horizontal
Payload	1 crew
Launch Mass	626 Mg (Titan 3C: 616 Mg, X-20: 10 Mg)
Propulsion System	Titan 3C: 5 x Rocket, X-20: 2 x Rocket
Status	Inactive, 1960 - 1963, Development (Canceled)
Image / Data Source	http://members.lycos.co.uk/spaceprojects/dynasoar.html
E-mail	spaceprojectsandinfo@aerospaceguide.every1.net

Dyna Soar means DYNAmic SOARing. Dyna Soar is a program to explore the questions of high altitude hypersonic flight.

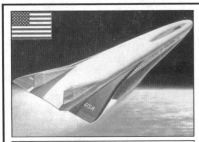

X-30 NASP	★☆☆☆☆
Developer	NASA
Type	Orbital
Number of Stages	1
Launch / Landing	Horizontal / Horizontal
Payload	2 pax
Launch Mass	140 Mg
Propulsion System	Ramjet + Scramjet
Status	Inactive, 1990 - n.a., Development (Canceled)
Image / Data Source	http://www.fas.org/irp/mystery/nasp.htm
E-mail	webmaster@fas.org

The X-30 was planned to be the demonstrator for the National Aero-Space Plane (NASP).

X-33	★★★★☆
Developer	NASA
Type	Suborbital
Number of Stages	1
Launch / Landing	Vertical / Horizontal
Payload	0 Mg (70 km)
Launch Mass	124 Mg
Propulsion System	2 x Rocket
Status	Inactive, 1996 - 1999, Development (Canceled)
Image / Data Source	http://trc.dfrc.nasa.gov
E-mail	carol.reukauf@dfrc.nasa.gov

The X-33 objective is to test novel "aerospike" rocket engines and metallic thermal protection systems.

USA

X-34 (Booster: L1011)	★★★☆☆
Developer	NASA
Type	Suborbital
Number of Stages	2 (PLN-r: 1, ORB-r: 1)
Launch / Landing	Horizontal / Horizontal
Payload	0,2 Mg (80 km)
Launch Mass	156 Mg (L1011: n.a., Orbiter: n.a.)
Propulsion System	L1011: 3 x Turbojet, X-34: 1 x Rocket
Status	Inactive, 1998 - 2000, Development (Canceled)
Image / Data Source	http://trc.dfrc.nasa.gov
E-mail	carol.reukauf@dfrc.nasa.gov

The X-34 is designed to demonstrate key vehicle and operational technologies applicable to future low-cost reusable launch vehicles.

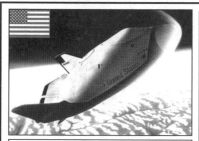

X-37 (Booster: Delta 4)	★★★☆☆
Developer	NASA
Type	Orbital
Number of Stages	n.a. (RKT-e: n.a., ORB-r: 1)
Launch / Landing	Vertical / Horizontal
Payload	0,5 Mg
Launch Mass	n.a. (Delta 4: 733 Mg, X-37: 4 Mg)
Propulsion System	Delta 4: n.a., Orbiter: n.a.
Status	Active, 1999 - Present, Development
Image / Data Source	http://trc.dfrc.nasa.gov
E-mail	carol.reukauf@dfrc.nasa.gov

Capable of being ferried into orbit on an expendable launch vehicle, the X-37 will test technologies during re-entry.

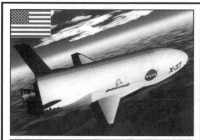

X-38 (Booster: Space Shuttle)	★★★☆☆
Developer	NASA
Type	Orbital
Number of Stages	3 (RKT-e: 1, FBB-r: 1, ORB-r: 1)
Launch / Landing	Vertical / Horizontal
Payload	6 pax
Launch Mass	2043 Mg (Space Shuttle: 2035 Mg, X-38: 8 Mg)
Propulsion System	Space Shuttle: 5 x Rocket, X-38: n.a.
Status	Inactive, 1995 - 2001, Development (Canceled)
Image / Data Source	http://trc.dfrc.nasa.gov
E-mail	carol.reukauf@dfrc.nasa.gov

The X-38 is a technology demonstrator for the proposed Crew Return Vehicle (CRV), which is designed as an emergency crew return vehicle from the ISS.

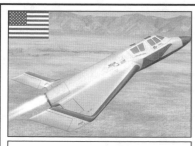

Xerus	★★★★☆
Developer	XCOR Aerospace
Type	Suborbital
Number of Stages	1
Launch / Landing	Horizontal / Horizontal
Payload	1 crew + 1 pax (100 km)
Launch Mass	n.a.
Propulsion System	1 x Rocket
Status	Active, 1999 - Present, Development
Image / Data Source	http://www.xcor.com
E-mail	webinfo@xcor.com

This vehicle has an option to carry payloads to LEO by attaching an expendable rocket on the back of the vehicle.

USA

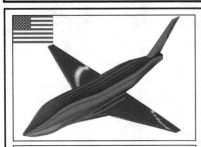

In computer simulations this vehicle is dropped from the booster at 48 km height and climbs until reaching engine burnout at approximately 60 km height.

XPV	★☆☆☆☆
Developer	Canyon Space Team
Type	Suborbital
Number of Stages	2
Launch / Landing	Horizontal / Horizontal
Payload	n.a. (100 km)
Launch Mass	n.a. (Booster: n.a., XPV: n.a.)
Propulsion System	Booster: n.a., XPV: Rocket
Status	Active, n.a. - Present, Development
Image / Data Source	http://www.canyonspaceteam.org
E-mail	info@canyonspaceteam.org

This vehicle looks like a slightly futuristic executive jet and uses a surplus J-85 jet engine to take it to an altitude of 11 km at Mach 0,8.

X Van 2001	★☆☆☆☆
Developer	Pan Aero
Type	Suborbital
Number of Stages	1
Launch / Landing	Horizontal / Horizontal
Payload	2 pax (100 km)
Launch Mass	3,6 Mg
Propulsion System	2 x Turbojet + 1 x Rocket
Status	Inactive, n.a. - n.a., n.a.
Image / Data Source	http://www.tour2space.com
E-mail	panaero@tour2space.com

References

[1] ANSI, 2000a   ANSI,   http://www.atis.org/tg2k/_leo.html,
                                          2000, accessed: 30/08/2005

[2] ANSI, 2000b   ANSI, http://www.atis.org/tg2k/_ geostationa-
                                          ry_orbit.html, 2000, accessed: 30/08/2005

[3] Arianespace   Shareholders,   http://www.arianespace.com,
                                          Evry-Courcouronnes,   France,   accessed:
                                          14/01/2005

[4] Chandler, A.D.   The Visible Hand: The Management Revolu-
                                          tion in American Business, Belknap Press,
                                          Cambridge, USA, 1979

[5] Chen, M.   Asian   Management   Systems,   ISBN
                                          1861529414, Thomson Learning, USA, 2004

[6] ESA   ESA,   http://www.esa.int/esaCP/index.html,
                                          accessed: 30/05/2004

[7] Goehlich, R.A.   A Representative Program Model for Devel-
                                          oping Space Tourism, ISBN 3-936846-29-4,
                                          Wissenschaftlicher Verlag Berlin, Germany,
                                          2003

[8] Harloff, G.J.   HASA - Hypersonic Aerospace Sizing Analy-
                                          sis for the preliminary Design of Aerospace
                                          Vehicles, USA, 1988

[9] Hobby Space   Hobby Space, http://www.hobbyspace.com,
                                          accessed: 30/09/2005

[10] HyperDic   Hyper Dictionary, http://www.hyperdictionary.
                                          com, accessed: 30/05/2005

[11] Ireland, J.   Finding the Right Management Approach,
                                          The China Business Review, Washington
                                          D.C., USA , 1991, pp. 14-17

[12] Koelle, D.E.

Handbook of Cost Engineering for Space Transportation Systems with Transcost 7.1, Ottobrunn, Germany, 2003

[13] Longsdon, J.M.

A sustainable Rationale for Manned Space Flight, Space Policy, Vol. 5, USA, 1989, pp. 3-6

[14] Moore, P.G.

The business of risk, ISBN 0-521-28497-X, Cambridge University Press, Cambridge, United Kingdom, 1983, pp. 97-115

[15] Ogasawara, Y.

A concept study on Horizontal Take-off / Landing Rocket Plane with Winged Tank, Master's Thesis, Keio University, Yokohama, Japan, 2004

[16] Ogawa, M.

A concept study on Horizontal Take-off / Landing Rocket, Bachelor Thesis, Keio University, Yokohama, Japan, 2003

[17] Olivier

Olivier Dubois-Marta, Launch Vehicles, 1994

[18] Space Notes

Japan Aerospace Exploration Agency, http://spaceinfo.jaxa.jp/note/note_e.html, Tokyo, Japan, accessed: 30/08/2005

[19] StarBooster

Starcraft Boosters Inc., http://www.star booster.com, accessed: 30/12/2004

[20] Terrenoire, P.

An Overview of the Reusable Rocket Launchers Based on Near Term Technologies, 45th International Astronautical Federation, Jerusalem, Israel, 1994

[21] X Prize

X Prize Foundation, http://www.xprize.org, St. Louis, USA, accessed: 30/07/2005

[22] Zwicky, F.

Entdecken, erfinden, forschen im morphologischen Weltbild, Droemer Knaur, München, Germany, 1966

Index of Spaceships